Skychart User Guide

A catalogue record for this book is available from the Hong Kong Public Libraries.

Published in Hong Kong by Samurai Media Limited.

Email: info@samuraimedia.org

ISBN 978-988-8406-80-7

Cartes du Ciel / Skychart

English documentation

Edited: October 03 2016

Last version is available from the wiki at
http://www.ap-i.net/skychart/en/documentation/start

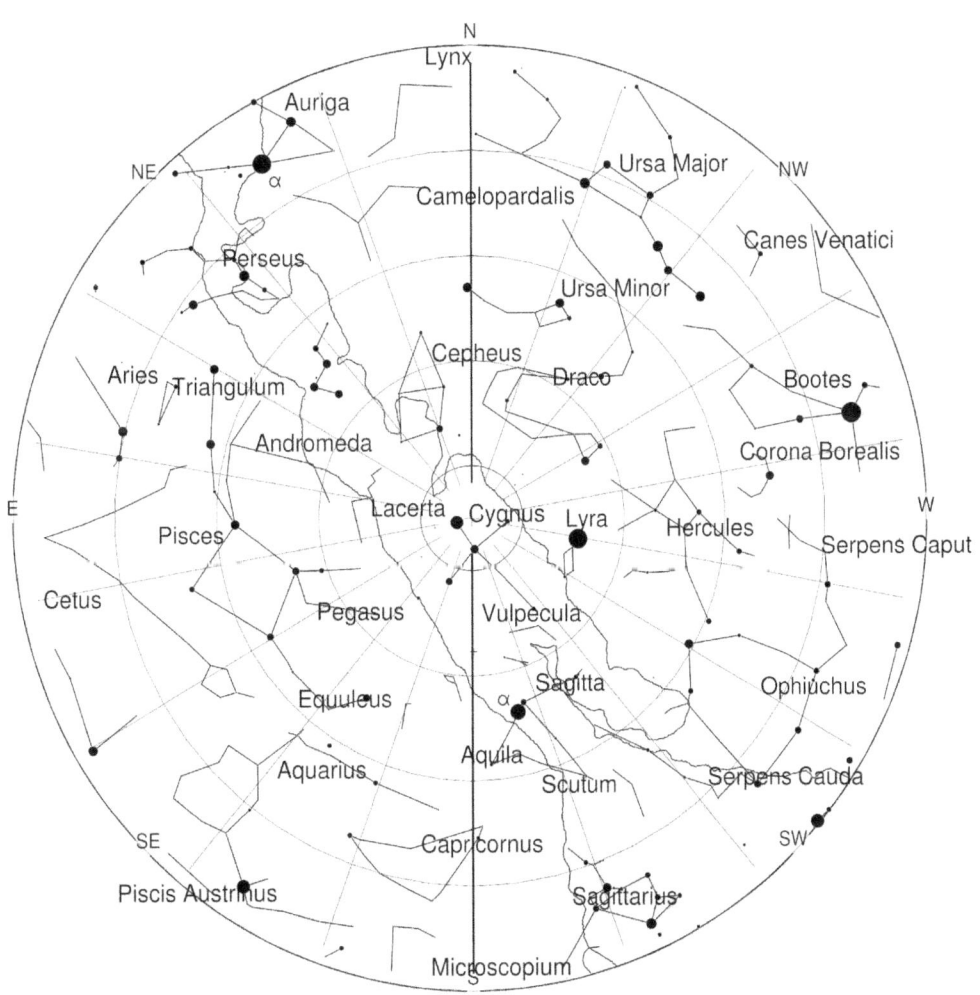

Documentation

You can contribute to these pages [1].

Download the PDF [http://www.ap-i.net/pub/skychart/doc/doc_en.pdf] version of this document.

Tutorial

- Quick Start Guide

Installation

- Installation on Windows
- Installation on Mac OS X
- Installation on Debian GNU/Linux
- Installation on Ubuntu
- Installation on Fedora
- Installation on Mageia
- Installation of extra catalogs
- Make a portable installation

Reference Manual

The menu bar and pop up windows from the chart

You can choose from the following menu items:

- File
- Edit
- Setup
- View
- Chart
- Telescope
- Window
- Update
- Help
- Pop-up windows from the chart

The tool bars

Since the version 3.11beta of May 2014 you can configure the tool bar button as you want by using the Tool bar editor. The "Standard" layout is described in the following pages.

- Main Bar
- Object Bar
- Left Bar
- Right Bar
- Left Tool box
- Tool bar editor

The Information Areas

- Status Bar
- Detailed Information
- Object List
- Information about the planets

The astronomical calendar

- Input Area
- Twilight
- Solar System
- Comet
- Asteroid
- Solar Eclipses
- Lunar Eclipses
- Artificial satellites

Settings

- Date, Time
- Observatory
- Chart, Coordinates
- Catalog
- Solar System
- Display
- Pictures
- General
- Internet

Tools

- Labels
- Advanced Search
- Position and Field of Vision
- Observing list
- SAMP Virtual observatory interface
- CatGen
- The Tool Box editor

Miscellaneous

- Add orbital elements manually
- Display of NEOs
- Keyboard shortcuts
- Command line options
- Server Commands
- Directories and Files
- Scripting reference
- Scripting examples
- Computation method and precision
- Installation and compilation of the source code
- FAQ

License

- Documentation license Creative Commons and GNU Free Documentation License
- Software license GNU General Public License

— *Patrick Chevalley [mailto:pch%20%5Bat%5D%20ap%20%5Bdash%5D%20i%20%5Bdot%5D%20net]* *2006/10/14 22:16*

---

1) by using the wiki at http://www.ap-i.net/skychart [http://www.ap-i.net/skychart]

Quick Start Guide

Memento download

Memento is the roadmap how to reach all the facilities in SkyChart / Cartes du Ciel.

To print this easily, the memento is available in an OpenDocument [http://en.wikipedia.org/wiki/OpenDocument] ('Open Office' document) or in PDF format.

memento_1.1.1_en.odt

memento_1.1.1_en.pdf

Guide Content

- **Installation**
- **The Chart**
- **Solar system**
- **Deep sky objects**
- **Driving a telescope**
- **Cartes du Ciel as an application server**

Installation

Download the binaries from **here [http://sourceforge.net/project/showfiles.php?group_id=64092&package_id=61207]** Launch the Skychart installer. If SkyChart V2 is already installed, you can use the same directory.

Advanced ⇒ for installation details on different platforms see:

- **Linux installation**
- **Mac installation**
- **Windows installation**

The Chart

Guide content

- **Installation**
- **The Chart**
- **The Solar System**
- **Deep Sky Objects**
- **Driving a telescope**
- **Cartes du Ciel-SkyChart as an application server**

When you start *Cartes du Ciel-SkyChart*, it will display something like this:

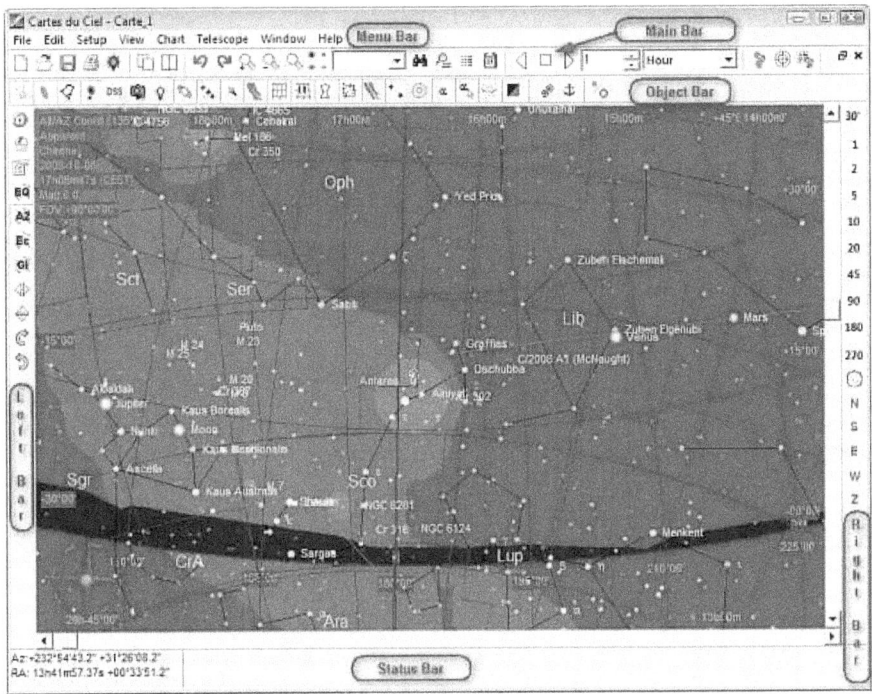

Observation location and time

With the **Setup → Observatory** dialog box (or with the 🔅 icon in left bar)you can choose a country, then a town close to your location.

Advanced ⇒ You can enlarge your choice by downloading more detailed locations and/or create new locations, see **Observatory**.

By default, when *Cartes du Ciel-SkyChart* starts the first time, the program will take your system time to calculate the object positions.
In **Setup → Date/Time** (or with the 🔅 icon in left bar), you can set your date/time of choice.
With the Time icons group ◁ ☐ ▷ ▷▷ 1 ↕ Hour ↕ in the main bar, you can modify the time step by step.

Advanced ⇒ You can simulate the movement of solar objects on the chart, see **Time simulation**

Lines / Grids

In the **Chart → Chart coordinate system** dialog box or with the **EQ**, **AZ**, **Ec**, **Gl** icons in the left bar you can choose between four coordinate systems:

- Equatorial
- Alt-Az (default)
- Ecliptic
- Galactic

For example, you can show the equatorial grid together with the Alt-Az grid on your map by **Chart → Lines/grids → Add Equatorial** grid or the 🔳 icon in the object bar.
The same object bar contains the following icons: 𝄖, 🔲, 🖊, ⋰ , these enable you to display constellations lines, constellation limits, the galactic equator and ecliptic.

You can also change the projection for the chart, see the projection comparison page.

Advanced ⇒ You can set more options in **Setup → Chart, coordinates**.

Horizon, Position

If you want to see sky in one of cardinal points, you can use **Chart → View Horizon** or one of the N, S,E, W icons in the right bar.

If you want to see the local zenith in the center of the chart, use the z icon in the right bar.

You can also use the ⚲ "Position" icon in the main bar, which opens a dialog box to enter coordinates for the center of the chart.

If you want to see the sky below the horizon, use **Chart → Below the horizon** or the icon ⟟ in the object bar.

Advanced ⇒ You can display a local horizon line, see the **Setup → Observatory → Horizon** dialog box.

Mirroring and rotating the chart

You can mirror horizontally, mirror vertically or rotate the chart right (clockwise) or left (counter clockwise) in steps of 15° at a time with **Chart → Transformation** or with the ⬖, ⬗, ↻, ↺ icons in the left bar.

If you want to rotate in steps of 1°, keep a shift key pressed while clicking on one of the rotation icons. The ⚲ "Position" icon in the main bar also gives you the posabillity to rotate the chart with a precision of a degree.

Zooming in, zooming out

The width of the chart is called Field of Vision (FOV) and is given in degrees. By default, eleven FOVs are defined in *Cartes du Ciel-SkyChart*.

You can modifiy the FOV with the:

- menu **Chart → Field of Vision**, followed by your choice.
- sensitive areas at the top of the bar at the right hand side.
- menu **View → Zoom in and Zoom out** or the +🔍 and -🔍 icons in the main bar which divide or multiply the FOV by two.
- 🔍 icon in the main bar which opens a dialog box for a cursor driven variation of the FOV.
- ⚲ icon in the main bar. The dialogbox enables you to set the center of the FOV with an accuracy of one second.
- ◯ icon in at the right hand side sets the FOV to 360°.

Advanced ⇒ You can modify the standard FOV ranges in the **Setup → Chart, coordinates → Field of Vision** dialog box.

Labels

You can switch on/off the display of labels from the menu by **Chart → Show labels** or with the α icon in the object bar.

With the α icon in the object bar you can switch "edit label" mode on. Then you can modify a label: when you right-click on it, a contextual menu is displayed to move it, change its contents or hide it.

Advanced ⇒ In the **Setup → Display → Labels** dialog box you can change fonts, font sizes and colors, choose the objects to be labeled, and if you wish, customize the standard contents.

Night Vision

To avoid a completely ruined night vision during your nocturnal observations, you can switch *Cartes du Ciel-SkyChart* to "Night Vision" (by the ⚙ icon in the main bar, or from the menu by **View → Night Vision**): The computer monitor will display a black background color and red shades for lines, labels and icons.

Search objects

You can search an object with the fast search input area [▾] 🔍 on the main bar. Enter its common name or entire identifier (with the catalog id).

The 🔍 icon opens an advanced search dialog box in which you are helped to choose the catalog.

When the object is found, it will be displayed at the center of the chart (but not locked).

Lock Charts

This facility places the selected object in the center of the chart until you unlock it. You can lock a chart on an object when you right-click on this object followed by a left click on the line "Lock on ..." in the **pop-up menu**, or use the ⚓ icon in the object bar.

Related to:

The **Auto-refresh every** check-box. (Setup → Date/Time, the Time tab) Locking and unlocking charts makes only sense when this checkbox is checked. Otherwise, the objects in your chart weren't moving at all.

Multiple Charts, Link

In *Cartes du Ciel-SkyChart* V3, you can open more than one chart in one instance of the application from the menu **File →
New Chart** or the ⬚ icon in the main bar.
You can arrange your charts from the menu bar by clicking on **Window** and making your choice in the pull-down window, or
by clicking the ⬚ ⬚ icons in the main bar.
Multiple charts can be linked, as a result all open charts will have the same coordinates for the center, regardless the scale
(FOV) or other settings for each chart. All charts inherit the coordinates from the chart that is active at the moment you set
the lock. To link the charts, use the ⬚ icon in the object bar. Or from the menu bar, click **Window**, and click on **Link all
chart** in the pull-down.

Fields of eyepiece and CCD

You can display marks on the chart which depict eyepieces or CCD fields with the **Chart → Lines/grids → Show mark** menu
or icon ⬚ in the object bar.
These marks are scaled with the chart.

Advanced ⇒ You are encouraged to describe your eyepieces and CCD in a dialogbox invoked by **Setup → Display → Finder
Circle and Finder rectangle**. A really handy feature to help you find your object.

Objects lists

You can choose the types of objects to be displayed on the chart by:

- **Chart → Show Objects** menu or
- by the ⬚ ⬚ ⬚ ⬚ ⬚ icons in the object bar or
- from the menu **Setup → Configure the program → Chart, coordinates → Object list**.

You can obtain a full list of the objects displayed on the chart with the ⬚ icon in the main bar. A window is opened, you can
sort the list by RA, print it or save it as a CSV file.

Detailed Information

When you select an object, the status bar on the bottom contains some information about it like the equatorial coordinates
and the name. You can obtain more detailed information by a right-click on the object. In the pop-up window, choose the first
entry **About ...**
In this detailed information window, you can choose to center the object or obtain a list of neighbouring objects (DE+/-1°, RA
+/- 4m). This list is an exerpt of the global objects list.

The fastest way to retrieve detailed information is to do a (left) click on the object of choice, and then a (left) click on the label.

Printing a chart

You can print the chart with the **File → Print** dialog box or by the ⬚ icon in the main bar.

Advanced ⇒ You can convert your chart in postscript or bitmap format with the **File → Print setup** dialog box.

Saving and Restoring Charts

You can save your chart in *Cartes du Ciel-SkyChart* format (to restore it later) with **File → Save as ...** or the ⬚ icon in the
main bar.
You can also save your chart in a PNG file with **File → Save image ....**

You can restore a saved chart in *Cartes du Ciel-SkyChart* format by **File → Open** or the ⬚ icon in the main bar.

The Solar System

Guide content

- **Installation**
- **The Chart**
- **The Solar System**
- **Deep Sky Objects**
- **Driving a telescope**
- **Cartes du Ciel-SkyChart as an application server**

Planets

You can switch on/off the display of the Sun, planets and satellites with the **Chart → Show Objects → Show planets** dialog box or the 🪐 icon in the object bar.

Advanced ⇒ In the **Setup → Solar system → Planet** dialog box, you can choose the appearance options for the Solar system objects, set the Jupiter GRS longitude, set on the Earth shadow (lunar eclipse) and a transparent mode of the planets (occulations).

Comets

The standard *Cartes du Ciel-SkyChart* installation contains a sample file of 207 comets that is loaded in the database.

Then you can switch on/off the display of comets with the **Chart → Show Objects → Show comets** dialog box or the 🔭 icon in the object bar.

Advanced ⇒ In **Setup → Solar system → Comet - Load MPC File** dialog box, you can load from MPC site a complete and fresh comets file. In the same dialog box, the tab "Data Maintenance" allows you to manage the comets database and "Add" enables you to add new comet caracteristics.

Tips: If you want to see all the comets in the file on a width angle chart set the following options: "Display as a symbol", … fainter than magnitude: 99, Show comets 99 magnitude fainter…

Asteroids

The standard *Cartes du Ciel-SkyChart* installation contains a sample file of 5000 asteroidsthat is loaded in the database.

You can switch on/off the display of asteroids with the **Chart → Show Objects → Show asteroids** dialog box or the ⬚ icon in the object bar.

Advanced ⇒ In **Setup → Solar system → Asteroid - Load MPC File** dialog box, you can load from the MPC site the complete and fresh asteroids file. Then you have to prepare the monthly data (same dialog box, next tab).
In the same menu, "Data Maintenance" allows you to manage the asteroids database and "Add" allows you to add new asteroid caracteristics.

Orbits simulation

You can simulate the movement of solar system objects on their orbits by the **Setup → Date/Time → Time simulation** dialog box.

Twilight simulation

You can simulate the twilight colour by the **Setup → Display → Sky colour** dialog box (radio button "Fixed Black" or "Automatic") or by the ▮ icon in the object bar.

Advanced ⇒ In the same dialog box, you can set twilight colour or fixed colour as you wish.

Ephemeris

You can obtain ephemeris for solar system objects in the **File → Calendar** dialog box or the 📅 icon in the main bar. This gives you information about twilight, planets, comets, asteroids, lunar and solar eclipses.

Advanced ⇒ In the same dialog box, you can download new solar and lunar eclipse predictions by Fred Espenak from NASA Eclipse Web Site [http://eclipse.gsfc.nasa.gov/eclipse.html].
You can also modify the colour of comets, asteroids and orbits in **Setup → Display → Display colour** and the twilight and sky background colour in **Setup → Display → Sky background colour**.

Deep Sky

Guide content

- Installation
- The Chart
- The Solar System
- **Deep Sky Objects**
- Driving a telescope
- Cartes du Ciel-SkyChart as an application server

Stars

You can switch on/off the display of stars with the **Chart → Show Objects → Show stars** dialog box or the ⚹ icon in the object bar.

You can modify the star appearance by the **Setup → Display Mode** dialog box or the ⚙ icon in the object bar (this modifies also the appearance of deep sky objects).

Advanced ⇒ You can find more options in the **Display Colour** tab of **Setup → Display** dialog box.

You can increase the amount of stars to be shown by the ⚹ icon, clicking it adds 0.5 to the maximum magnitude limit. To decrease the number of displayed stars, click the ⚹ icon to decrease the maximum magnitude limit by 0.5. These icons are positioned in main bar.

Advanced ⇒ You can set the maximum magnitude limit for stars by the FOV ranges in **Setup → Chart, coordinates → Object Filter**.
You can show the proper motion of stars by **Setup → Lines**.

Deep Sky Objects

You can switch on/off the display of deep sky objects with the **Chart → Show Objects → Show Deep Sky Objects** dialog or the ⚹ icon in the object bar.

Nebulae outlines (such as M42 in Orion) can be display with the **Chart → Show Objects → Show Lines** dialog box or the ⚹ icon in the object bar.

The appearance of feep sky objects can be modified by the **Setup → Display Mode** dialog box or the ⚙ icon in the object bar (this modifies also the appearance of the stars.)

Advanced ⇒ You can find more options in the **Display Colour**, **Deep sky object Colour** tabs of **Setup → Display** menu.

You can increase the amount of deep sky objects to be displayed by the ⚹ icon, clicking it adds 0.5 to maximum magnitude limit. Clicking the ⚹ icon decreases the maximum magnitude limit by 0.5. These icons are positioned in the main bar.

Advanced ⇒ You can select the display of deep sky objects by maximum magnitude limit and maximum size in arcminutes by FOV ranges in **Setup → Chart, coordinates → Object Filter**.

Catalogs

The standard *Cartes du Ciel-SkyChart* installation comes with:

- **Bright Stars Catalog** which contains 9096 stars to a maximum magnitude of 6.5 (461 k).
- **SAC** which contains 10,600 deep sky objects (2.6 MB)

Advanced ⇒ Many other catalogs can be dowloaded from **SourceForge //Cartes du Ciel-SkyChart// Catalog Download Page [http://sourceforge.net/project/showfiles.php?group_id=64092&package_id=208104]**. Some of the supplementary catalogs overlap or objects can't be drawed with a large FOV, so you can choose catalogs to activate according to your needs in **Setup → Catalog** menu (**CdC Stars** and **CdC Deep Sky** tabs).

For details about downloading extra catalogs, see Installation of extra catalogs.
For configuration of extra catalogs in *Cartes du Ciel-SkyChart*, see Catalog

Pictures

To obtain a more realistic appearance of deep sky objects rather than symbols, you have to download from **pictures** an excerpt of RealSky pictures corresponding to the SAC catalog content and install it. This is not bundled with *Cartes du Ciel-SkyChart*, because of it size.
Then, you have to build the pictures database with **Setup → Pictures → Object** tab (scan directory).

Many of the pictures are sized around 1°x1°, so they are only visible with a FOV below 30°. You can switch the display of pictures between on or off by the **Chart → Show Objects → Show pictures** dialog box or the 🔆 icon on the object bar.

DSS/RealSky

For any particular deep sky object, you can download a picture from the **DSS Site [http://archive.eso.org/dss/dss]** by the ᴅss icon in the object bar.
It is temporary stored in **Your_profile\documents\Cartes du Ciel\pictures\$temp.fit** as defined in the **Setup → Pictures DSS/RealSky** tab (temporary file).
You can choose to display these pictures by the 🖼 icon in the object bar.

Advanced ⇒ You can set more options in the **Setup → Pictures → DSS RealSky** dialog box.

Driving a telescope

Guide content

- **Installation**
- **The Chart**
- **The Solar System**
- **Deep Sky Objects**
- **Driving a telescope**
- **Cartes du Ciel-SkyChart as an application server**

The basic steps are:

1. Select the telescope driver (ASCOM or other driver; if ASCOM or INDI, be sure to have installed the corresponding drivers)
2. Once you have selected the telescope driver, go to telescope control panel and set the basic telescope configuration: COM port, model variant, ...
3. When ready (all cables set, hand-pad initialization done), connect the telescope using the **Connect** button.
4. Click the **Slew** button to command the telescope to go to your target.

The complete documentation is available with the Telescope Menu

See specific the help about each driver for ASCOM, INDI, LX200 and Encoder.

Cartes du Ciel-SkyChart as an application server

Guide content

- **Installation**
- **The Chart**
- **The Solar System**
- **Deep Sky Objects**
- **Driving a telescope**
- ***Cartes du Ciel-SkyChart* as an application server**

It is possible to use Skychart as a server to automate some task.

This is the function that is used when another software can open Skychart to show a map of the object you are working on. This functions are normally documented in the calling software.

On the Skychart side there is only minimal configuration available. Just be sure that **Use TCP/IP Server** is checked, that **Server IP interface** is set to 127.0.0.1 and the **Server IP port** is 3292. A restart of the program is required after a change in this options.

If you plan to connect to Skychart from another computer on your network you need to change the **Server IP interface** to 0.0.0.0

To know if a program is connected to Skychart you can use the Server Information window.

You cannot exit Skychart as long as another application is connected. In this case all the way to close Skychart just make it to minimize in the task bar.

To close Skychart you need to quit all the connected application first. Depending on the application it may also close Skychart automatically.

You can develop your own applications or scripts to automate Skychart. See the list of Server Commands and example scripts.

Installation on Windows

Instructions to install the binaries on a Windows system

It is possible to install the *Cartes du Ciel-SkyChart* version 3.0 over a full Windows version 2.76 in the same directory, and run the *Cartes du Ciel-SkyChart* 2.76 version apart from the 3.0 version. But you can also install the 3.0 version of Cartes du Ciel-SkyChart only.

Run the installer file `skychart_3.xxxxx.exe`, to install Cartes du Ciel-SkyChart in your prefered directory (i.e. C:\Program Files\Ciel)

At any time you can start the version 3 with `skychart.exe`, or the 2.76 version with `ciel.exe`

Options

Telescope drivers

Only a limited number of telescopes are natively supported.
To work with the full range of telescope models you need to install the ASCOM platform [http://ascom-standards.org/] and the required telescope drivers [http://ascom-standards.org/Downloads/ScopeDrivers.htm].

Video recording

Video recording require the ffmpeg [http://www.ffmpeg.org] software.

You can get a ready to install version for Windows from a number of places:
http://www.videohelp.com/tools/ffmpeg [http://www.videohelp.com/tools/ffmpeg]
http://sourceforge.net/projects/mplayer-win32/files/FFmpeg/ [http://sourceforge.net/projects/mplayer-win32/files/FFmpeg/]
http://ffmpeg.arrozcru.org/autobuilds/ [http://ffmpeg.arrozcru.org/autobuilds/]

You can install the files anywhere you want, but be sure to set the location of ffmpeg.exe in the menu Setup / Date-Time / Animation.

Installation on Mac OS X

Prerequisites

- A recent Mac with Intel processor, PPC processors are not supported at the moment.
- Mac OS X 10.6 or better.

Instructions to install the binaries on Mac OS X.

- Download the install file, i.e. skychart-3.1-i386-macosx.dmg
- Open the file to mount the virtual disk.
- The disk folder opens, it contains the package installer skychart.pkg.
- Right click the package icon, select Open and follow the instruction to install the software. Depending on your security setting you may have to accept that the package is unsigned. Read more information about Gatekeeper [http://support.apple.com/kb/HT5290]
- You can now umount the virtual disk (drag to the trash).

Start the program

- Open the Applications - Cartes du Ciel folder.
- Open the "skychart" icon.

Options

Telescope drivers

The INDI [http://indi.sourceforge.net] drivers are required to drive the telescope. Look at this page [http://www.indilib.org/index.php?title=Devices] to know if your telescope is supported.

The simplest way is to install a package for Mac OS X available from http://www.cloudmakers.eu/indi [http://www.cloudmakers.eu/indi]

You can also install INDI from source using MacPort [http://www.macports.org].

- If not already done, install Xcode from your Mac distribution CDrom or dowload it from Apple web site.
- Install MacPort using the .dmg file for your Mac OS version available from http://www.macports.org/install.php [http://www.macports.org/install.php]
- Install INDI from a terminal with this command:

  ```
  sudo port install indi
  ```

- To update to a new version use this command:

  ```
  sudo port selfupdate
  sudo port upgrade outdated
  ```

Wine

Wine is required to compute the artificial satellites position in the Calendar.

The easiest way is to install a pre-build package available from https://dl.winehq.org/wine-builds/macosx [https://dl.winehq.org/wine-builds/macosx]
See http://wiki.winehq.org/MacOSX [http://wiki.winehq.org/MacOSX]

You can also install from MacPort [http://www.macports.org] as for INDI. The most tricky part is to make the new PATH environment know from the applications launched from the Application menu.

- Do the first two step above to install Xcode and MacPort.
- Open a terminal and type:

  ```
  sudo port install wine
  ```

- After the installation logout and login again, open a terminal and type

  ```
  wine --version
  ```

  to check it is installed successfully.

- Type the following command to make it available from Skychart:

```
defaults write ${HOME}/.MacOSX/environment PATH "${PATH}"
```

- Logout and login again before to try with Skychart.
- If it still say that wine is not installed, type the command:

```
echo $PATH
```

Copy the result and edit the file /etc/launchd.conf to add the following line including the full content of your $PATH environment:

```
setenv PATH /opt/local/bin:/opt/local/sbin:...
```

DOSBox

If you also want the Iridium flare prediction with the satellites pass you need to install the DOSBox software.

- Download the DOSBox dmg from: http://www.dosbox.com/download.php?main=1 [http://www.dosbox.com/download.php?main=1]
- Install in /Applications
- To let Skychart use it, copy the binary file to a location in you PATH, for example the same location as Wine above. Be sure to rename the file with all lowercase letter when you do the copy:

```
sudo cp /Applications/DOSBox.app/Contents/MacOS/DOSBox /opt/local/bin/dosbox
```

- To test, type dosbox from a terminal.

Video recording

Video recording requires the ffmpeg [http://www.ffmpeg.org] software.

You can install it with MacPort [http://www.macports.org]

- Install MacPort as indicated above for INDI.
- Install ffmpeg from a terminal with this command:

```
sudo port install ffmpeg
```

A static version of the program is also available here: http://ffmpeg.arrozcru.org/autobuilds/ [http://ffmpeg.arrozcru.org/autobuilds/]

You can also try to install it this way [http://hints.macworld.com/article.php?story=20061220082125312].

Installation on Debian GNU/Linux

The Debian packages are available to enable automatic installation using apt-get. Supported architecture are i386, amd64, armhf.

First, open a root shell or use sudo.

Add Patrick Chevalley's signature to APT:

```
apt-key adv --keyserver keyserver.ubuntu.com --recv-keys D79BF92A
```

Add a new source file in `/etc/apt/sources.list.d`
For the stable version do:

```
echo "deb http://www.ap-i.net/apt stable main" > /etc/apt/sources.list.d/skychart.list
```

If you want more frequent update with the development version do:

```
echo "deb http://www.ap-i.net/apt unstable main" > /etc/apt/sources.list.d/skychart.list
```

Then refresh your package list, from a command line type:

```
apt-get update
```

To install the software with all the prerequisite and the additional data files:

```
apt-get install skychart
```

The prerequisite software are normally automatically installed. If not, check you have the packages libpasastro, libgtk2.0-0, libglib2.0-0, libpango1.0-0, libjpeg62, libpng12-0, libsqlite3-0, xplanet, indi, ffmpeg

That's all, exit the root shell and type skychart to run the program.

The update for a new version is done at the same time as the standard update for the other software of your distribution, normally once a week on Monday for the unstable release.

Other software you can install from this same repository are:

```
ccdciel
eqmodgui
indistarter
```

Installation on Ubuntu

Bellow installation procedure uses only command line approach, because this is the simpler way to follow instructions just by copy-pasting the commands in terminal. Instructions tested on Ubuntu 12.04, 11.10 and 11.04.

1. Add Skychart repository:

   ```
   sudo apt-add-repository 'deb http://www.ap-i.net/apt stable main'
   ```

   Note: If you would like to install development version of Skychart instead of stable version, then replace word "stable" with "unstable" in above and bellow command.

2. Above command also adds source code repository 'deb-src http://www.ap-i.net/apt [http://www.ap-i.net/apt] stable main' which should not be added in order the program to work at all, so removing source code repository:

   ```
   sudo apt-add-repository --remove 'deb-src http://www.ap-i.net/apt stable main'
   ```

3. Request public key:

   ```
   gpg --keyserver keyserver.ubuntu.com --recv C56CCB02D79BF92A
   ```

4. Add public key:

   ```
   gpg --export --armor C56CCB02D79BF92A | sudo apt-key add -
   ```

5. Update repository:

   ```
   sudo apt-get update
   ```

6. Install Skychart without full dependencies (does not install the packages required for the Artificial Satellites display, you can install them later):

   ```
   sudo apt-get install --no-install-recommends skychart
   ```

7. This is all software you need to install if sky will be observed with naked eye. But if you like to see some deep-sky objects (e.g. galaxies) and stars that are not seen with naked eye then install additional packages (this will download several MB of additional software):

   ```
   sudo apt-get install skychart-data-stars skychart-data-dso skychart-data-pictures
   ```

8. Launch Skychart from terminal:

   ```
   skychart
   ```

   You will also find the Skychart button in the main menu, using "Science & Engineering" filter.
   Note: If installed following above instructions, 'skychart' command is actually a script that disables some Ubuntu Unity features (only for Skychart program) to make it possible Skychart to work correctly (LIBOVERLAY_SCROLLBAR=0 and UBUNTU_MENUPROXY=0).

9. The program starts up, it takes several seconds to initialize, click Next button.
10. Observatory window opens. You need to tell the program from which location you will be observing the sky. You can select your observation place by clicking on Observatory database and select the city you are observing from. If the city is not included on the observatory database list you can download more details for your country, or you can manually enter Latitude, Longitude and Altitude information from the main Observation window. Search this information from e.g. web page: http://www.findlatitudeandlongitude.com/ [http://www.findlatitudeandlongitude.com/]
11. That is it. Program starts. Try to click one of the buttons on the right sight N (North), S (South), E (East) or W (West) to get sky image on screen like it is in the nature. See the quick start guide for more information about the different use of the program.

Installation on Linux Fedora

Base software

The stable version of Skychart is included in the Fedora packages.

To install the software, just search for skychart in the Add/Remove Software menu:

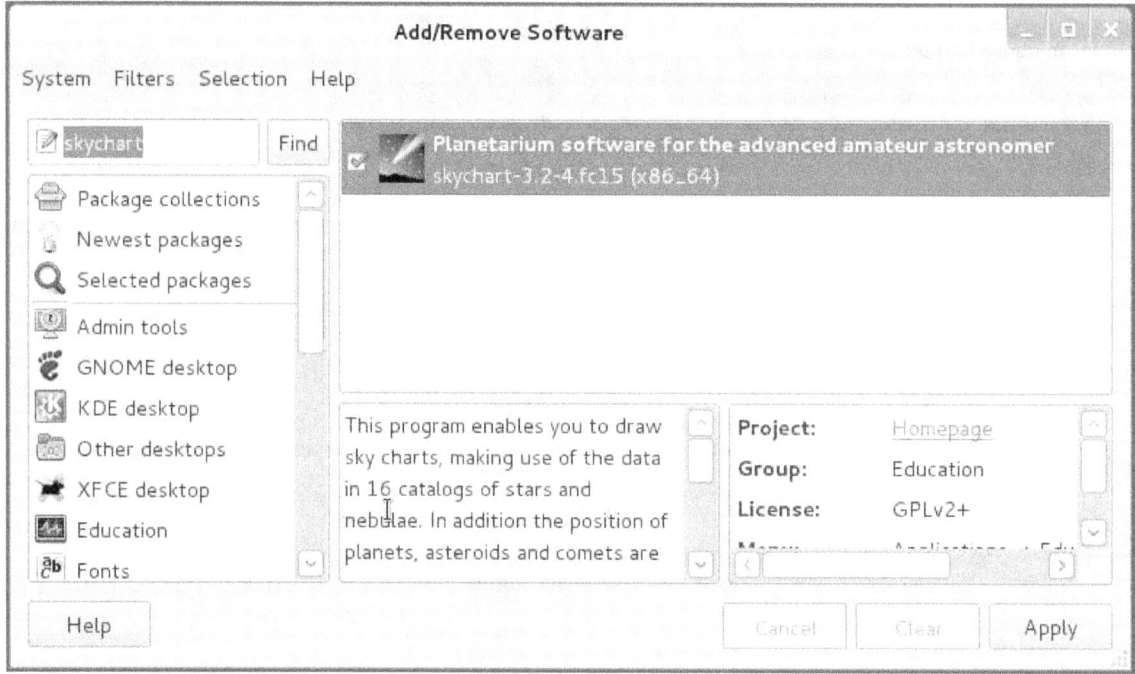

To install from a command line terminal, type as root:

```
yum install skychart
```

Please note that, due to Fedora policies, some functionalities of Skychart are not available in the rpm provided by Fedora repositories (artificial satellites computation and Iridium flares prediction).
If you need such functionalities please use rpm provided on Skychart homepage.

Documentation

To save bandwith, starting with Skychart 3.6, offline documentation is provided by a separate package named skychart-doc. To install it from a command line terminal, type as root:

```
yum install skychart-doc
```

Additional Catalog

Only minimal data are include with software base package.
For more stars or DSO objects you can install the additional data packages:

```
yum install skychart-data-stars
```

and / or

```
yum install skychart-data-dso
```

For additional pictures look at the download page.

Installation on Linux Mageia

Base software

The stable version of Skychart is already included into the packages provided by the Mageia repositories.

To install the software, just use rpmdrake (or using the Mageia Control Center) as shown in the figure below:

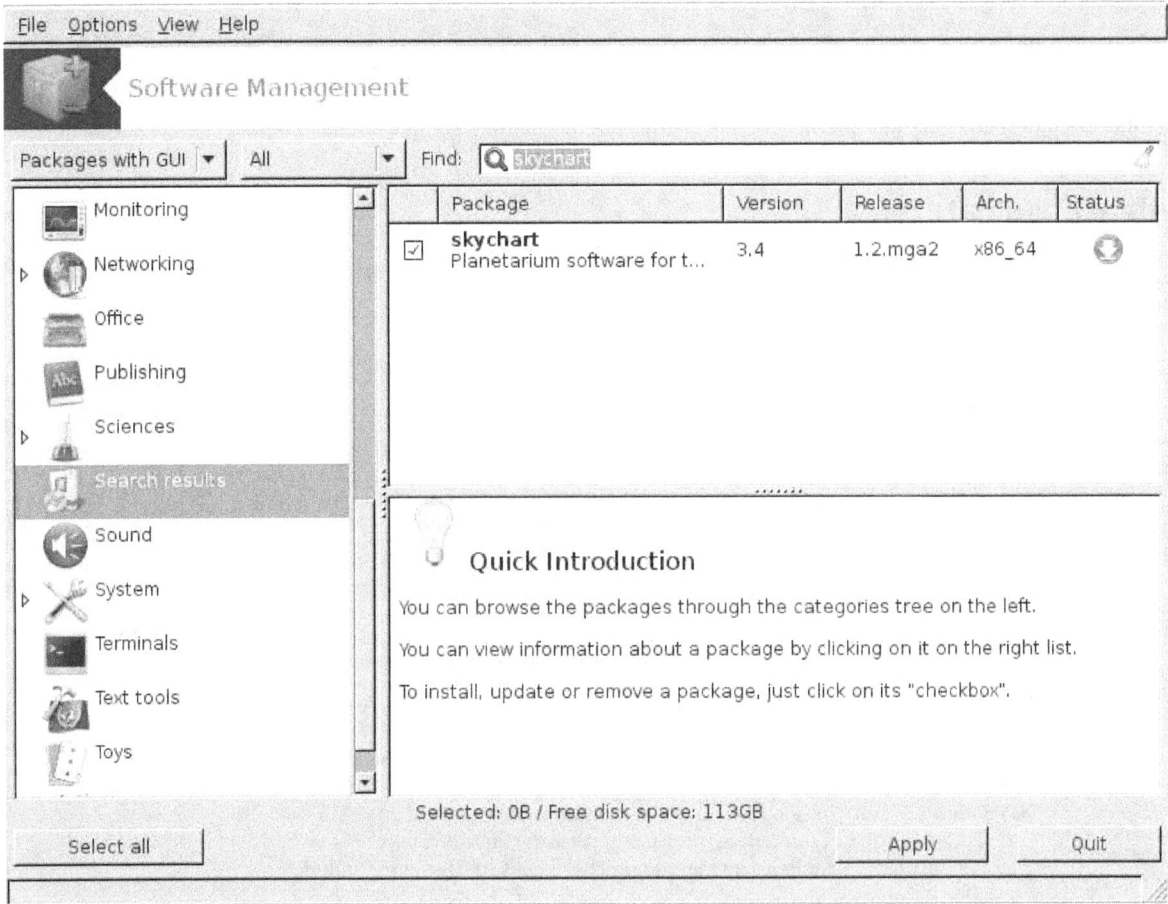

To install from a command line terminal, type as root:

```
urpmi skychart
```

or using sudo as a standard user:

```
sudo urpmi skychart
```

Some functionalities of Skychart are not available in the rpm provided by Mageia repositories (artificial satellites computation and Iridium flares prediction).
If you need such functionalities please use rpm provided on Skychart homepage.

Documentation

To save bandwith, starting with Skychart 3.4, offline documentation is provided by a separate package named `skychart-wikidoc`. To install it from a command line terminal, type as root:

```
urpmi skychart-wikidoc
```

Additional Catalog

Only minimal data are include with software package.
You can get RPM files for more stars, nebulae and pictures on the download page.

Installation of extra catalogs

With *Cartes du Ciel-SkyChart* it is easy to install popular ready-to-use catalogs. You can download and use them for free. Most catalogs aren't larger than some tens of megabytes, these will be helpfull to many dedicated amateurs.

But maybe you want to push things a little further. For example, you want your charts to display stars to a magnitude of 19. Then you need to download the USNO-A2.0 catalog. Before you start downloading, consider its size. Installed, this catalog will consume 6.11 GB. Still interested? See **The large catalogs**

You also can choose to build your own catalog. As a basis you can choose from thousands of existing catalogs and modify them for usage in *Cartes du Ciel-SkyChart*. You also can gather information in your own ASCII-file. To use these kinds of catalogs, you need the tool **CatGen** to adapt them for *Cartes du Ciel-SkyChart*. Those catalogs demand a little more time before you can use them in *Cartes du Ciel-SkyChart*. But this is a chapter about the ready-to-use catalogs, so lets continue with that:

The ready to use catalogs

To get to the source of those catalogs you go to the **download link [http://www.ap-i.net/skychart/en/download/]** from the **Cartes du Ciel-SkyChart homepage [http://www.ap-i.net/skychart/]**.

Catalog name	Description	installation directory	installed size
cdc_base_catalog	Base catalog. An absolute necessity. Usually you get this delivered with the program installer for stable or beta versions. It contains: **XHIP** the Extended Hipparcos Star Catalog **SAC** the Saguaro Astronomy Club version 8.1 Deep Sky objects catalog with index **Deep Sky Outlines** outlines of the bright nebulae **Index** the NGC, Messier and IC search-index files	cat/xhip cat/DSoutlines cat/ngc2000 cat/sac	total 24 MB
cdc additional stars catalog	This add the following star catalog: **Tycho-2** star catalog, containing data about 2.5 million stars to magnitude 11. **WDS** the Washington Double Star Catalog, contains data of astrometric multiple star systems. **GCVS** variable stars catalog with information about all kinds of variable stars. You can find here eruptive, pulsating, rotating, cataclysmic, eclipsing, intense variable X-ray stars and other types further subdivided in this one catalog. **Search Index** for SAO, BD, HD, GC star number.	cat/tycho2 cat/gcvs cat/wds	total 85 MB
UCAC 4 catalog	This add the following star catalog: **UCAC 4** star catalog, include 113 millions stars to magnitude 16. Read more information here	cat/ucac4	total 5.9 GB
cdc additional nebulae catalog	A group of catalogs of Deep Sky Objects. **GCM**: Globular Clusters in the Milky Way (Harris, 1999) contains data of 147 globular clusters close to our Milky Way. **GPN**: Catalogue of Galactic Planetary Nebulae (Acker+, 1992). 1143 proven and probable planetary nebulae, and 347 possible planetary nebulae. **LBN**: Lynds' Catalogue of Bright Nebulae (Lynds 1965). Also contains a cross-reference to NGC, Index Catalogue (IC), Sharpless (1959) Catalogue of HII regions, Cederblad (1956) Catalogue of Diffuse Galactic Nebulae, and Dorschner and Gurtler (1963). **NGC2000**: This is a modernized collection of the New General Catalogue of Nebulae and Clusters of Stars (NGC), the Index Catalogue (IC), and the Second Index Catalogue composed by J. L. E. Dreyer (1888, 1895, 1908). Contains 13.226 Deep Sky objects, equinox B2000.0 **OCL**: The fifth edition of the Lund Catalogue of Open Cluster Data, provides key information about all known open clusters in our Milky Way. **PGC** Catalogue of Principal Galaxies, extract from the 2012 Hyperleda database, contains data about 1.5 million galaxies.	cat/gcm cat/gpn cat/lbn cat/ngc2000 cat/ocl cat/pgc	total 174 MB
cdc_rngc_catalog.zip	By Wolfgang Steinicke reviewed and by **Jeff Burton [http://x.astrogeek.org/software/cdc/catalog.php]** for *Cartes du Ciel-SkyChart* adapted list of NGC and IC objects, originally composed by Dreyer. Contains exactly data about 14.000 Deep Sky objects. Equinox here is 2000.0, this version dates from november 24th 2002. Activate this catalog in the first tab of the **catalog settings** dialog box by setting	cat/RNGC	2,54 MB

	the path to the ".hdr" file.			

When you want to install these catalogs for all users, you need to have *Administrator* (Windows) or *root* (Linux) rights. If you don't have these, you can choose to install the catalogs somewhere in your computer where you have sufficient rights. (For example, your home directory.)

Download the catalog files that you want, save them (temporary) somewhere on your computer. Use your favorite unzip-program to decompress the files to the **installation directory** of *Cartes du Ciel-SkyChart*. With Windows this is typically something like **C:\Program Files\Ciel**, with Linux ususally this is **/usr/share/skychart**. But you can choose another path. Remember to configure the path to your catalog with **Setup → Catalog**.

Activation of the catalogs

You can activate or deactivate your catalogs with the dialog boxes from **Setup → Catalog**. Click the appropriate tab for your specific catalog.

The large catalogs

HST GSC original FITS

HST-GSC is an acronym for "Hubble Space Telescope Guide Star Catalogue". The original goal was to keep the HST properly pointed at its target. This catalog contains over 19 million objects brighter then magnitude 16, of which 15 million were identified as stars. This catalog is considered obsolete. Only to be complete, I 'll describe how to retrieve this catalog. I recommend you to use the **HST GSC Compact**.

This large catalog is available from **ftp://adc.gsfc.nasa.gov/pub/adc/superseded/1/1220/GSC/** **[ftp://adc.gsfc.nasa.gov/pub/adc/superseded/1/1220/GSC/]**. After you have done everything that is needed to use this catalog, you will discover that this catalog consumes 1.18 GB of space. The .gsc files in this version contain ASCII data.

In the directory of the ftp-server you find *directories* and 'tarball' *files* with names like LFFF.tar.gz. (Yes, there is also a file with the name N0730.tar.) To download this catalog, you can choose to do it in a fast way or a slow way.
The fast way: download all tarball files to your computer, including the N0730.tar. To correct this small error, rename N0730.tar to N0730.tar.gz.
Next step is to 'untar' all tarballs with your favorite archive (unzip) program. (With Windows, I use **ZipGenius** **[http://www.zipgenius.com/]**. Every Linux distribution knows how to deal with tarballs.) Usually, the content of the tarball will be written in a directory with the same name as the tarball with omission of the '.tar.gz' part. In this directory you use your favorite archive program again to decompress every file with the **.gz** extension.

Watch this: After the untar action, pretty often the contents of the tarballs (i.e. N0000.tar.gz) ends up in a directory export\pub\ftp\pub\adc\archives\superseded\1\1220\GSC\LFFFF, and pretty often they don't. Whatever the result, after decompression move all 'LFFFF' directories with their content to one common directory. After that, set *Cartes du Ciel-SkyChart* to use the common path with the **Setup → Catalog → CDC Stars** dialog box.

Downloading the tarball is quite fast, the decompression of all the .gz files however isn't. (It's well possible that it will take you a few hours to do them all.) In any case, this is the fast way. The slow way would be to download every single .gz file from every subdirectory of the FTP-server. And then you still need to decompress them..

HST GSC Compact

From the *Cartes du Ciel-SkyChart* point of view there is not much difference between the *HST-GSC original FITS* and the *HST GSC Compact*. Again, it's based on the data of 19 million objects brighter than magnitude 16, of which 15 million are identified as stars. The size of the HST GSC Compact makes all the difference: on your harddisk it only needs 290 MB. This is because the .gsc files in this catalog are in a binary format.
The original *Cartes du Ciel-SkyChart* version 2.7x allready could work with the **version 1.1 [ftp://cdsarc.u-strasbg.fr/pub/cats/I/220/GSC/]**. Now, this catalog is considered obsolete. The **version 1.2 [ftp://cdsarc.u-strasbg.fr/pub/cats/I/254/GSC/]** version became its successor (also obsolete now), at this moment we recommend you the HST-ACT version to serve as your HST GSC Compact catalog. You can download this one from

- **ftp://cdsarc.u-strasbg.fr/pub/cats/I/255/GSC_ACT/ [ftp://cdsarc.u-strasbg.fr/pub/cats/I/255/GSC_ACT/]**, or from
- **ftp://adc.astro.umd.edu/pub/adc/archives/catalogs/1/1255/GSC_ACT/** **[ftp://adc.astro.umd.edu/pub/adc/archives/catalogs/1/1255/GSC_ACT/]**.

When your only possibility is to use HTTP, you might consider downloading from

- **http://cdsarc.u-strasbg.fr/ftp/cats/i/255/GSC_ACT/ [http://cdsarc.u-strasbg.fr/ftp/cats/i/255/GSC_ACT/]**.

From your download source, copy all files to a local directory. If you still have got the older HST GSC compact versions 1.1 en 1.2, you still can use them with *Cartes du Ciel-SkyChart*.

Last but not least: Activate your catalog with the **Setup → Catalog → CDC Stars** dialog box.

USNO-A2.0

As I said earlier, this is a large catalog. It contains data of 526,280,881 stars, among the data are the magnitudes in V and B. The limiting magnitude is at 19+. Unfortunatly, this catalog doesn't contain data about the proper motion.

There are some versions of the 'USNO A' catalogs, the most recent is USNO A version 2.0. There is also a smaller subset from the 'USNO A' catalog. This summary is considerable smaller and is called USNO-SA 2.0, but it is useless to plot a chart.

And then there is the USNO B version. Since the arrival of this catalog, the professionals consider USNO A2.0 as obsolete. USNO B contains data of 1 billion objects, complete with data about their proper motions. See the next paragraph.

The installed catalog will occupy 6,11 GB on your media.
The 'USNO A' catalog is no more available from NOFS so I put a copy for you on my own web server.
To download the USNO-A2.0 catalog, use the following link to copy everything from

- **https://vega.ap-i.net/pub/usnoa/ [https://vega.ap-i.net/pub/usnoa/]**

Wherever you retrieve your USNO-A catalog from, **make sure** that (after eventually decompression) your catalog contains files with a *.acc* **and** *.cat extensions*. Files from ftp://cdsarc.u-strasbg.fr/pub/cats/I/252/USNO_A2 [ftp://cdsarc.u-strasbg.fr/pub/cats/I/252/USNO_A2] contain files with a *.ppm* extension. I didn't get these to work with *Cartes du Ciel-SkyChart*, save yourself from this dissapointment.

Copy files one by one. If you don't do that, you 'll risk that your process will be killed at the server site, simply because your task allocated to much memory. (These files are large, remember?) After downloading, put all the files together in one directory if you didn't do that allready.

And again: Activate your catalog with the **Setup → Catalog → CDC Stars** dialog box

USNO-B1.0

Since the version 3.9 Skychart can use the USNO-B1.0 catalog in U.S. Naval Observatory format, 180 directory 000 to 179, each with 10 .acc and 10 .cat files, 78 GB total size.

You can download this catalog using Bittorent [http://en.wikipedia.org/wiki/BitTorrent_(protocol)], download size is 46 GB:

- Get the torrent file [http://www.ap-i.net/pub/skychart/usnob/usno-b1.0.torrent]
- Open the torrent file in your bittorrent client software [http://en.wikipedia.org/wiki/Comparison_of_BitTorrent_clients]. You can eventually select only the zone directory you want. Each zip file contain the catalog data for a one degree declination zone, the name of the file is the South Polar Distance (DEC+90) of the zone.
- When the download is complete unzip each of the 180 zip files to recreate the original catalog structure.
- Please keep seeding to help other to get the data!

The configuration setting is in the **Catalog → Obsolete** tab for two reason: to not bore the many people that cannot get this files, but also because I really think that NOMAD or PPMXL is a real improvement over this catalog.

NOMAD

The Naval Observatory Merged Astrometric Dataset (NOMAD) contains astrometric and photometric data for over 1 billion stars derived from the Hipparcos , Tycho-2 , UCAC2 , and USNO-B1.0 catalogs for astrometry and optical photometry, supplemented by 2MASS near-infrared photometry.

See on this page the NOMAD installation detail with Skychart. The data itself need to be downloaded using Bittorrent.

PPMXL

The PPMXL catalog of positions and proper motions on the ICRS. Combining USNO-B1.0 and the two Micron All Sky Survey (2MASS).
It contain about 900 million stars, is complete down to magnitude 20 and include the proper motion for every object.

See on this page the PPMXL installation detail with Skychart. The full data is easily to download from the CDS.

Make a portable installation of Skychart

A portable install mean you install Skychart on a removable media (USB stick, external disk) and you can plug it to any PC to run the application without any further configuration. The configuration is also written to the removable media so you retrieve your preferred settings on any computer.
The version 3.8 or more recent is required for this process

The example is given here for Windows but you can do almost the same on Linux, in this case use the tar installer for an easy setup.
You can also use the portable version for Windows on Linux/Mac by using Wine. This way you need a single USB stick for any platform.

The commands suppose your removable media is mounted on the drive E: change accordingly if it use another letter.

This description use mostly a command line window for the clarity of the explanation but you can also use the equivalent graphical tool.
To open a command line window use the Start menu → Accessories → Command Prompt

1) Create a new folder on your removable media:

```
E:
mkdir portable_skychart
cd \portable_skychart
mkdir Ciel
```

2) Download the Skychart **Windows zip** from the Download page. Save the zip file in the folder **E:\portable_skychart\Ciel**.

3) Extract the zip file in this folder:

```
cd \portable_skychart\Ciel
unzip skychart-3.8-2450-windows.zip
```

4) Create a folder for the program configuration:

```
cd \portable_skychart
mkdir userdata
```

5) Create a startup script:

```
cd \portable_skychart
notepad skychart.cmd
```

Copy the following lines and save the file:

```
@ECHO off
set basedir=%CD%
start %basedir%\Ciel\skychart.exe --config="%basedir%\userdata\skychart.ini" --userdir="%basedir%\userdata"
```

Now you can plug this USB stick on any (Windows) computer and run the program with a double click on skychart.cmd.

Optional steps

Do not let any trace in the registry

If you not want to let any trace in the registry of the computer you need to inactivate the server functionality. Otherwise a registry key indicating the connection port is created.

```
cd userdata
notepad skychart.ini
```

Locate the line starting with AutostartServer, set the following and save the file:

```
[main]
AutostartServer=0
```

Copy the program setting

Since the version 3.8 it is possible to copy the configuration file skychart.ini from an existing installation to the userdata folder.

Another option is to save and reload a chart using the menu File/Save as , File/Open.

Make a script to run on Linux with Wine

Mount the configured USB key on your Linux system.

```
cd /media/my-usb-key/portable_skychart
vi skychart.sh
```

Copy the following lines and save the file:

```
#!/bin/bash
wine cmd /C skychart.cmd
```

As you cannot set the executable bit on a FAT file system do the following to run:

```
cd /media/my-usb-key/portable_skychart
bash skychart.sh
```

File Menu

New Chart

The new V3 of `Cartes du Ciel` enables you to open different windows with separate sky charts. The advantage is that you can visualize different charts at the same moment for different locations, time or even display configuration.

After creating a new chart you can arrange the charts through the Window menu. You can maximize, minimize, close or even resize each chart by dragging the right and bottom edges.

Open

Use this to open a previously saved sky chart with its own configuration of the location and time.

Save As...

This saves the active sky chart to a file that you can reload in the future with the Open option.

Close Chart

This will close the active chart, only when you have more than one chart open.

Save Image...

The active sky chart is saved as an image. It is possible to select PNG, JPEG or BMP format for the file.

Print

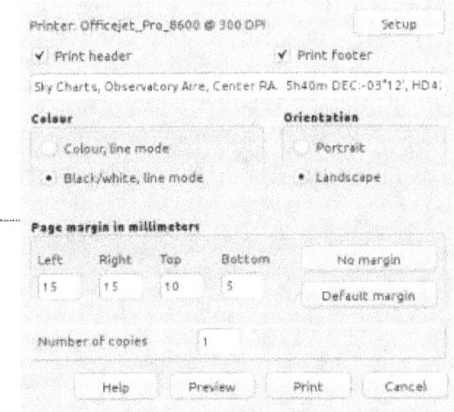

The Print dialog box enables you to choose your settings for the 'printer-target', regardless if this is a real printer or a file.

You can select to print a header with the description text, and a footer with the chart legend.

Printing can be done in the colours as on the screen or in black&white. The orientation, the paper margins and number of copies can also be selected.
If the printer is set to .bmp file is also possible to choose printing method to reversed black&white (with black background sky).

The default orientation is Landscape to better match the screen format and get a print result similar to what is displayed on the screen. If you want to print in Portrait orientation it is recommended to resize the screen window to this format to adjust the framing and objects visibility.

Print Preview

This option is only visible when you select a real printer in Printer Setup below.

It show a preview of the printed output using the current settings in order to save some paper when you adjust this options.

Printer Setup

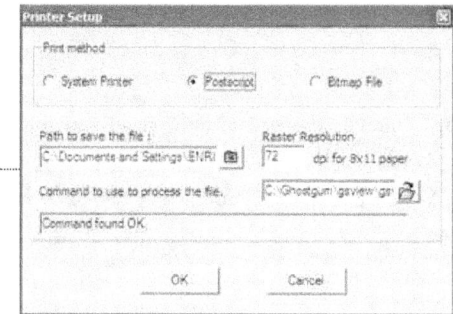

Configures the printer to print the screen content. There are three options:

1. System Printer: It enables you to configure a real printer.
2. Postscript: It enables you print to a postscript file. This is the way to go if you want a vector file you can further convert to PDF or SVG. On Windows it requires you to have Ghostscript and GsView32 installed and a correctly configured path to view the result. It will only work with version 7.xx of GsView32.
3. Bitmap file: It enables you to print the chart to a BMP file. It is required to have a correctly configured path to the MsPaint executable.

Exit

Just that, exits the program Cartes du Ciel

Edit Menu

The **Edit** menu consists of the following items:

- Advanced Search
- Edit Label
- Copy
- Undo
- Redo

Advanced Search

This brings you to the dialog box to do an advanced search for an object. You can search for planets, stars, deep sky objects, comets or asteroids. Click here to read more about the Advanced Search dialog box.

Edit Label

If you want to edit labels, the Edit Label mode needs to be set to editable, and that's just what a click on this line is all about. Every click on this line inverts the Edit Label mode, So, in fact it doesn't change the content of any label. You also can change the Edit Label mode by a click on the ⍺ icon on the objects bar.

To read how to modify a label, read about the modification of labels.

Copy

Copies the currently active chart into the system clipboard, so you can easily past it as an image in a word processor or imaging program.

Undo

This will undo the last action executed in the program.

Redo

This will redo the last action that was previously cancelled by an Undo action.

Setup Menu

Tool bar editor

The Tool bar editor allow you to edit the button bar layout.

Manage tool box script

Select to function you want to assign to each of the left tool box panel.

Date/Time

Gives you control over the used time in the chart. Here you also can make your settings to simulate the movement of objects in the solar system. A click the line **Date/Time** opens the configuration window. Have a look at **this** page for further details.

Observatory

This enables you to configure the position of the observer and how to display the horizon. Click on the line **Observatory** to open these configuration dialog boxes. Look at **this** page for further details.

Chart, Coordinates

Lets you choose from several possibilities to display your chart, such as the coordinate system, field of views and the grid spacing. A click here opens the **Chart, Coordinates** configuration dialog boxes. Look at **this** page for further details.

Catalog

Here you can choose which catalogs the use to display objects on the chart. Click here to open the **Catalog** configuration window. Look at **this** page for further details.

Solar System

Lets you set planet rendering options and manage the comet and asteroid data. Click this line to open the **Solar System** configuration dialogs. Look at **this** page for further details.

Display

Here you can customize the appearance of the graphical interface. Configure things like the sky colour, lines, labels and fonts of the program. Also, if you want to mark the field of vision of your eyepieces or CCD/CMOS sensors on the chart, configure them here. Click the line here to go to the first tab of the **Display** configuration dialog boxes. Look at **this** page for further details.

Pictures

Here you can configure settings how to combine the display of object pictures on the chart. Here you also can set to you use RealSky (Local installation, if you've got the CD's) or the online DSS services. Click this line to go to the **Pictures** configuration window. Look at **this** page for further details.

General

Miscellaneous settings about the database and telescope connections, and the setting of the language in the program. Click here to open the **General** configuration window. Look at **this** page for further details.

Internet

Lets you configure how your computer connects to the Internet and set the links to the online resources. Click this line to go to the **Internet** configuration dialog boxes. Look at **this** page for further details.

All configuration options

Opens the main configuration window from where you can make all program settings.
This window is divided into several sections, every section contains the same possibilities as described on the previous paragraphs of this page. So here are the shortcuts:

- **date_time**
- **Observatory**

- **chart coordinates**
- **Catalog**
- **Solar System**
- **Display**
- **Pictures**
- **System**
- **Internet**

Apply Change to all Chart

By checking the small checkbox "Apply change to all chart" at the bottom of the configuration window, you apply all the configuration changes to all your opened charts. When this box is unchecked, the configuration changes will only be applied to the active chart.

Save Configuration Now

Clicking on this will cause the program to save the current settings as the new default. A newly opened chart will use these settings. Also, when you click **File → Reset chart and options**, the active chart will be set according to these saved settings.

Save Configuration On Exit

If you check this line, the program will ask you to save the settings every time you exit the program.

Reset language

Reset the language to the default value for your computer.
For evident reason this menu text is not translatable and always show in English.

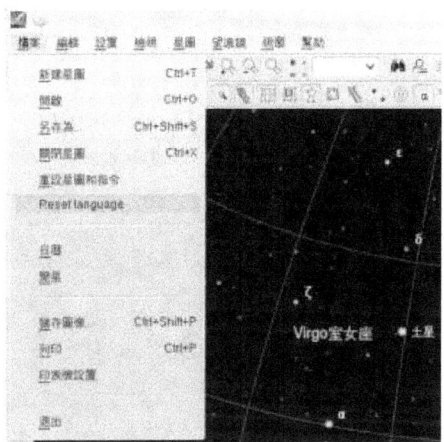

Reset chart and options

This give you three options:

- **Reset initial default and restart** This delete all your personal settings and restart the program as the first time you run it.
- **Reset to last time the chart was saved** Reset the chart and the options to the last saved configuration. The result is the same as when you would close the map without saving followed by restarting the program.
- **Set options for best performance** Reset every option that affect the program performance. The visual aspect of the chart is simplified and the performance must be acceptable on every computer.

View Menu

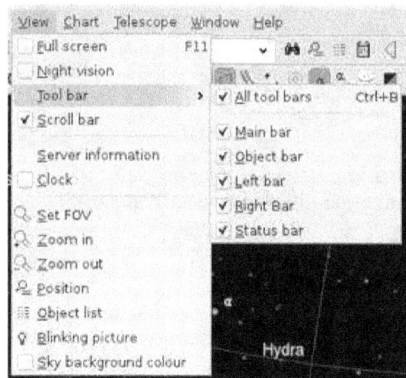

Full Screen

From the menu: **View → Full Screen**

This resizes the CdC window to fill the entire screen. The display status of the tool bars, status bar, scroll bars and the menu bar will be unchanged. A new click on this entry resizes the CdC window to its original size.

Pressing the **F11** key is a shortcut for this function.

Night Vision

From the menu: **View → Night Vision**

Click on this entry to swap the colour of the chart between a normal display and a night vision display:
black background, red shades for grids, labels and other lines. The icons on the tool bar will change as is defined in the "themes", at the bottom right of the Setup → Display → **Color** dialog box.
With Windows Vista, the desktop background colour becomes middle grey.
You can also change to and from night vision by clicking the ☀ icon on the **main bar**.

Windows specifics

If your Windows system use the "modern" XP, Vista or Aero theme some part of the screen like button, scroll bar or menu cannot change their color.
If you need a true dark display for your observation change for the "Windows Classic" theme on your computer, or disable the visual theme for skychart.exe as indicated on the figure to the right. Then switch to full screen mode with the F11 key to eliminate the title bar.

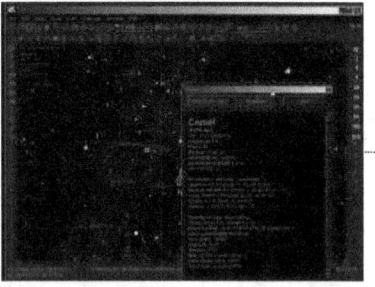

Linux specifics

The widget color cannot be changed by an application but are the result of the choice of a Gtk2 theme.
There is so many dark Gtk2 theme available that it can make the choice difficult. Be sure to close and restart skychart after you try a new theme.
Then select Nightvision to change the chart color.

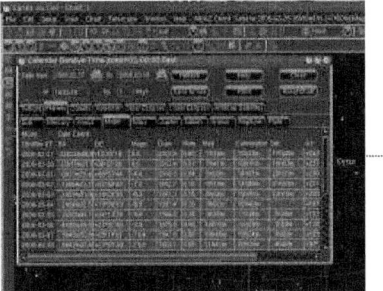

Tool Bar

From the menu: **View → Tool Bar**

This shows a secondary menu with these entries:

- **All Tools bar** enable or disable the display of all bars, except the menu bar (**shortcut Ctrl+B**).
- **Main bar** enable or disable the display of the horizontal bar under the menu bar, above the chart.
- **Object bar** enable or disable the display of the horizontal bar under the main bar, above the chart.
- **Left bar** enable or disable the display of the vertical bar at left of the chart.
- **Right bar** enable or disable the display of the vertical bar at right of the chart.
- **Status bar** enable or disable the display of the horizontal bar below the chart.

Tool Box

Select the tool box you want to activate on the left of the screen. You can also use the Function keys F1 to F8.

Scroll Bar

From the menu: **View → Scroll Bar**

This enables or disables the display of scroll bars to shift the chart, on the bottom and at the right. This makes movement of the chart possible, by dragging the marker over the scroll bar. You also can click on the scroll bar somewhere near the marker to shift the chart in large steps, or click on the small arrows at the end of the scrollbar to move the chart in small steps.

You also can move the chart without the scroll bars by keeping a shift-key pressed while pressing the left mouse-key and moving the mouse. You can also move the chart by pressing the up, down, left and right keys with or without while keeping the Ctrl, Shift or Alt keys pressed. For these possibilities, see keyboard shortcuts.

Server Information

From the menu: **View → Server Information**

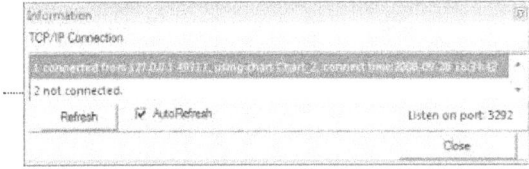

This shows you the connection status of the Skychart Server TCP/IP with its clients. You can refresh it or choose "auto refresh".

You can also close a connection by a click with your right mouse key on a line.

SAMP

SAMP is a messaging protocol that enables astronomy software tools to interoperate and communicate. Skychart can send and receive coordinates position, FITS images and VO tables.

The submenu include this functions:

- **Connect to SAMP hub** : Connect Skychart to a running SAMP hub, for example Topcat [http://www.star.bris.ac.uk/~mbt/topcat/] or Aladin [http://aladin.u-strasbg.fr/aladin.gml] include such hub.
- **Disconnect from SAMP** : Disconnect from the SAMP hub.
- **SAMP status** : Display the status of the connection.
- **SAMP setup** : Open the SAMP setup window.

Variable Stars

Launch the Variable Stars Observer program.

Clock

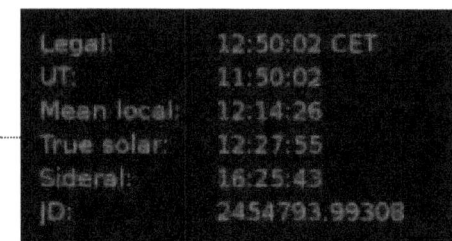

From the menu: **View → Clock**.

This shows a window with the current time information. It lists the following items:

- The legal time for your **time zone [http://en.wikipedia.org/wiki/Time_zone]**.
- The **Universal Time [http://en.wikipedia.org/wiki/Universal_Time]**.
- The **Mean local solar time [http://en.wikipedia.org/wiki/Solar_time#Mean_solar_time]** for your longitude.
- The **True solar time [http://en.wikipedia.org/wiki/Solar_time#Apparent_solar_time]**, based on the apparent Sun hour angle.
- The local **sidereal time [http://en.wikipedia.org/wiki/Sidereal_time]**.
- The **Julian day number [http://en.wikipedia.org/wiki/Julian_day]** (UT)

This clock came as a separate program, called **cdcicon**. After installing the SkyCharts package, the clock can be run standalone from the SkysCharts installation directory. Administators under Windows can run it from the system tray too.

If you want to change the time used by SkyCharts, click the 🜨 icon from the **left bar**, or from the menu **Setup → Date/Time**.

Solar system information

Open a window that give information about the planets, their visibility and their orbits.

The visibility chart also include the currently selected object.

Calendar

This function informs you about some astronomical phenomenona for a given time period.

The window is composed of seven areas:

- an **input** area. Look at this page for further details.
- a **Twilight** tab. Look at this page for further details.
- a **Planet** tab. Look at this page for further details.
- a **Comet** tab. Look at this page for further details.
- a **Asteroid** tab. Look at this page for further details.
- a **Solar Eclipses** tab. Look at this page for further details.
- a **Lunar Eclipses** tab. Look at this page for further details.

Observing list

Open the Observing list window.

Trajectories simulation

Open the planet trajectories simulation window.

Change Mouse mode

Same function as the Change Mouse mode button.

Distance measurement

Same function as the Distance measurement button.

Set FOV

From the menu: **View → Set FOV**

This makes a continuous variation of the Field of Vision (FOV) possible. After a click on this line, a small dialog window pops up, with a marker on a logarithmical scale. Simply click with the left mouse key on the marker, and while keeping the mouse key pressed, shift the marker by moving the mouse to set the FOV as you wish.

This is the same function as you get when you click the 🔍 icon from the **zoom group view** icons on the main bar.

If you want to set a very precise FOV, try the FOV possibility from **View → Position**.

Zoom In

From the menu: **View → Zoom in**

By a click on this line, the current FOV will be devided by two.

The +🔍 icon in the **main bar** is a shortcut for this function. You can also modify the FOV by turning the mouse wheel. See for more options about changing the FOV the **keyboard shortcuts**.

Zoom Out

From the menu: **View → Zoom Out**

This multiplies the current FOV by two.

The -🔍 icon in the **main bar** is a shortcut. You can also modify the FOV by turning the mouse wheel. See for more options about changing the FOV the **keyboard shortcuts**.

Position

From the menu: **View → position**

After a click on this line a dialog box pops up, here you can read or set the position parameters of the center of the active chart. The position can be in every kind of coordination grid. Here you also can read or set the parameters for Field of Vision (FOV) or the rotation angle of the chart. Click **here** to read all the details.

A click on the 🔍 icon from the **main bar** gives you the same dialog box.

Object List

From the menu: **View → Object List**.
By a click on this line you will retrieve a list of the displayed objects in the chart. You can filter the object types to list by configuring them in the tab **Object List**, from the menu: **Setup → Chart, Coordinates**.
You also can retrieve this object list by a click on the ⊞ icon in the **main bar**.

Image List

If some image are displayed this open the Image List window.

Blinking Picture

From the menu: **View → Blinking Picture**.
You can do the same by clicking the apropriate icon from the objects bar from the **pictures group**.

When you previously loaded a FITS-picture, you can 'blink' this picture with the original chart by a click on this line.
If you didn't load a picture, clicking this line doesn't change anything.

To read more about how to open a FITS-picture from a local source, click **here**.
To read more about how to download a picture from the online Digital Sky Survey (DSS), click **here**.

To read more about the display of these images, click **here**.
To read more about the configuration of the DSS resources to download those images, press **here**.

Sky Background Color

From the menu: **View → Sky Background Color**.

When you have set the chart to use the Alt-Az coordination grid and you have set the sky to automatic display of the sky background colour, it is possible that the displayed sky background colour isn't very dark.
This is the case when the configured position and time is such that the Sun isn't more than 18º below the horizon (daytime, or still no astronomical darkness reached, or the Moon above the horizon.). In this case, by every click on this line you can switch the displayed sky background colour between the "fixed sky colour" (usually black) or the colour that was generated according to the "automatic" setting.

You can do the same by clicking the ▰ icon from the **marks group** in the object bar.

To change between an automatic or fixed sky background colour, click **setup → Display**, then click the **Sky Color** tab.

If you didn't set the "Sky Color" to automatic **and** didn't set the chart to use the Alt-Az coordination grid the sky background colour will be fixed. Under these conditions, clicking this line will not change anything.

The Chart Menu

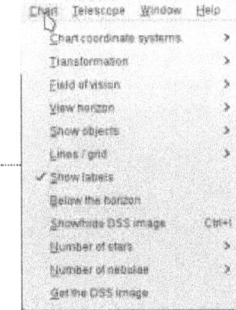

The Chart Menu enables you to configure the displayed chart according to your needs in a fast and easy way. If a more detailed configuration is needed, take a look at **Setup → Chart, Coordinates**

Chart Coordinate System

From the menu: **Chart → Chart Coordinate Sytem → ....**

Here you can choose from

- Equatorial Coordinates,
- Azimuthal Coordinates,
- Ecliptic Coordinates
- Galactic Coordinates.

These functions are identical with the **coordinate system group** of icons at the left hand side bar.
With **Setup → Chart, Coordinates → Chart coordinate system** you can configure the equinox and epoch as wel.

Transformation

From the menu: **Chart → Transformation →**

If you want to mirror or rotate the chart, you can make those settings with this submenu by clicking the appropriate line. Rotating the chart by this way goes in steps of 15º of arc.

A click on one of the corresponding icons from the **transfomation group** in the lower part of the tool bar at the left has the same result as a click on a line of this menu. You can rotate in steps of 1º by a click on this icon while pressing the shift key.

Field of Vision

From the menu: **Chart → Field of Vision →**

Here you can set the width/hight of the displayed field of vision of your active chart to a predefined value (in degrees of arc).

This works identical with the upper **field of vision group** icons at the right hand side tool bar.
You can change the default settings for the fields of Vision according to your preferences by **Setup → Chart, coordinates → Field of Vision**.

View Horizon

From the menu: **Chart → View Horizon → [Direction]**

If you are interested in objects situated in one of the cardinal directions, simply click the appropriate line to make your choice:

- North,
- South,
- West,

- East.

Setting the chart by these methods to one of the cardinal directions, also forces the chart to use the Altitude-Azimuth coordinate system.

This works identical with the lower **horizon group** icons at the right hand side tool bar.
To read about how to display your local horizon, click **here**.

Animation

This is the same function as in the main button bar

Show Objects

From the menu: **Chart → Show objects → Show ...**
By every click on one of these lines you check or uncheck the display of an entire category of objects.
In this submenu you can decide which of the object types you want to display on the chart by checking or unchecking the listed items in the submenu. These items are:

- Show Stars
- Show Deep Sky Objects
- Show Pictures
- Show Lines
- Show Planets
- Show Asteroids
- Show Comets
- Show Milky Way

This works identical with the icons from the **Object Group A** or **Object Group B** at the left of the object tool bar, on top of the chart. These items can also be found in the **Display lines** and **Solar system** tabs.

It is important to have set the right **date/time** and **place of observation** for a proper display of these items on your chart.

Lines - Grid

From the menu: **Chart → Lines / Grid →**

When you want to activate or deactivate the display of some grids or lines on the chart, you can do this by checking or unchecking the listed items of this submenu:

- Show Coordinate Grid
- Add Equatorial Grid
- Show Constellation Line
- Show Constellation Limit
- Show Galactic Equator
- Show Ecliptic
- Show Mark

This works identical with the ⊞ 🔲 ⚹ 🔲 ✏ *. and ⊙ icons in the **object tool bar**, on top of the chart.

You can configure the displayed type of lines in the tab **Lines** from **Setup → Display**. You can configure the displayed dimensions of the eyepiece and CCD in the tabs **Finder circle (Eyepiece)** and **Display rectangle (CCD)** from **Setup → Display**.
You can configure the grid spacing from the menu by **Setup → Chart, Coordinates → the Grid Spacing** tab.

Show Labels

From the menu: **Chart → Show Labels**

- α Activate the labels, as the button in the **Marks Group** of the object tool bar.
- Abc Activate the chart information display in the chart top left area.
- ▦ Activate the chart legend.

Activating this item will show labels for all objects according to the configuration from **Setup → Display → Labels**.

Below the horizon

From the menu: **Chart → Below the Horizon**

When you activate this, the chart will display you all of the sky as if the Earth and everything is transparent. This way you can view objects which are below the horizon for the configured observatory and the time set. By a click on the line (check or uncheck) from the menu or icon you switch this behaviour on or of. Only available when your chart is set to Alt-Az coordination grid.

This works identical with the ☄ icon in the **marks group** of the object tool bar.

Show-Hide DSS image

From the menu: **Chart → Show/Hide DSS image**

If you previously loaded a FITS formatted image, you can switch its display on or off by a click on this line. An interesting feature is the ♀ blink icon, to 'blink' the loaded image with the original chart content.

To read more about how to load a FITS picture, click **here**.
To read more about the configuration of the usage of DSS, click **here**.
To read more about the configuration of the DSS resources to download those images, press **here**.

The only shortcut to show or hide the image is to press **CTRL-I**.

Number of Stars

From the menu: **Chart → Number of Stars → [More/Less] Stars**
This works identical with the ● or · icons in the **magnitude group** of the main bar.

Here you can choose to increase or decrease the number of stars and Solar system objects by changing the displayed magnitude limit by 0.5.

Related to:

Setup → Chart, Coordinates; the Object Filter tab

When the *Filter Stars* checkbox is unchecked, the stars magnitude group buttons lose their function.

Number of nebulae

From the menu: **Chart → Number of Nebulae → [More/Less] Deep Sky**
This works identical with the ● or · icons in the **magnitude group** of the main bar.

Here you can choose to increase or decrease the number of deep sky objects by changing the displayed magnitude limit by 0.5.

Related to:

Setup → Chart, Coordinates; the Object Filter tab

When the *Filter Deep Sky Objects* checkbox is unchecked, the deep sky magnitude group buttons lose their function.

Get DSS Image

From the menu: **Chart → Get DSS Image**

This works identical with the DSS icon in the **pictures group** of the object bar.

You can load FITS formatted pictures from RealSky, the online Digital Sky Survey (DSS) site, from your SAC pictures catalog [program installation directory/data/pictures/sac or any other source.
Most users will use this feature to load a FITS-picture from the Digital Sky Survey (DSS) site. The size of pictures that you request are limited by the field of vision. Usually, you can't retrieve pictures when you set SkyCharts to a FOV larger than two degrees of arc. The larger the FOV, the greater chances are that your request times out, or isn't supported by the server at all. You need to realize that it takes a lot of CPU power at the server site to generate a picture, you have to be patient.

You can skip to push the download button every time you want to retrieve data from online resources. To do so, you need to uncheck the checkbox "Ask confirmation before any Internet connection" in the first tab of the dialog box, retrieved from the menu by **Setup → Internet**.

To read more about the display of these images, click **here**.
To read more about the configuration of the usage of RealSky and DSS, click **here**.
To read more about the configuration of the DSS resources to download those images, press **here**.

Telescope Menu

From the menu: **Telescope**

SkyChart can be used to drive electronically controlled mounts, but you can also use SkyChart to retrieve instructions to turn knobs on a manually controlled mount. Anyhow, before you can use your telescope mount in combination with SkyChart, you need to configure the sort of mount and the driver from **Setup → System → Telescope**.

Telescope settings

Same as **Setup → System → Telescope**.

Coordinates system

It is very important that both the program and the telescope driver use the same coordinate system.

Most INDI and ASCOM drivers specify if they want J2000 or Local coordinates and Skychart obey to that. Even for Local coordinates this is always non refracted coordinates.

Control Panel

From the menu: **Telescope → Control Panel**

Here, you can make specific settings for the driver that you chose earlier in the dialog. There are a only a few drivers but they are usable with a multitude of mounts with each their specific possibilities. Please, check the manual that came with the driver and mount.

See specific help about each panel for ASCOM, INDI, LX200 and Encoder.

A very important feature that all driver specific dialogs share, is the **Connect** button. When the telescope is physically connected and the driver/mount specific settings are OK, the program usually connects automatically. The connection status is displayed at the bottom of the dialog. Somewhere between the `Connect` and the `Disconnect` button, there is a coloured square. Red indicates a disconnected status, green indicates the connected status.

As soon as your telescope is connected to your mount, SkyChart will immediately read the current coordinates from your mount into your chart. After that, SkyChart will show you this position centered on the chart.

The ⬦ icon of the **telescope group** on the main bar is a shortcut for this function.

INDI Control Panel

This option is only show when the telescope interface is INDI

Open a standard INDI option setting panel for all your INDI devices

Slew

From the menu: **Telescope → Slew**

After your telescope mount is **connected** and **synchronized** to an object, you can slew your telescope to another object. Select the desired object simply by a click. Now, from the menu, click on **Telescope**, in the pull down, click on the line **Slew**. Now the coordinates are send to the mount. In many cases, the mount will start slewing immediately. On some mount controllers it might be neccesary to confirm your settings.

The slew ⬦ icon of the **telescope group** on the main bar is a shortcut for this function.
Another way to slew your telescope to the next object, is by a right mouse button click on the label of the selected object. In the pop up window, click the line **Telescope**, and in the next pop up **Slew**.

You can stop the current slew operation with the **Abort Slew** menu or the keys: **Ctrl+K**.

Sync

From the menu: **Telescope → Sync**

This is the way to enter the coordinates of a selected object on the chart to your telescope mount. And of course, in order to do this, your telescope must be connected to the computer. The connection must be physically by the correct cables and interface, and also logically by the program as well.
It is easy: first of all, you direct your telescope to an object that you positively identified. The next step is to select this object

on the chart, just a click on the object is enough. Now, click from the menu **Telescope** on the line **Sync**. Now, the coordinates of the selected object on the chart are entered into your mount. From now on, the mount knows the coordinates of its position at the sky at which it is pointed.
When you 've set your **markers** on, the markers will be shown in white, with the selected object in the center.

In the **telescope group** on the main bar you find the ⊕ sync icon as a shortcut for this fuction.
Another way to sync your telescope to the chart, is by clicking the label of a selected object with the right mouse button. In the pop up window, click the line **Telescope**, and in the next pop up **Sync**.

Track telescope

Center the chart on the telescope position and follow it.

It is important to remember that a telescope connection is linked to the chart that was active when you connect it.
This allow to use more than one telescope, even with a different driver, each one connected to it's chart.

So if you want many chart to follow a single telescope do it in the chart with the narrower field of view, and link the other chart to this one.

Window Menu

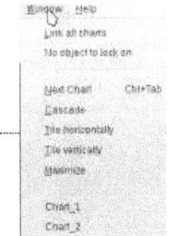

Settings of the position and behaviour of the charts inside a SkyChart window.

Link all Chart

From the menu: **Window → Link All chart**

You can have more charts open with different fields of vision, rotations or different coordination grids set. When you activate to link all charts, all charts will copy the center position of the active chart. Every change of the center position in the active chart will be followed by the other charts. You still can change the FOV or orientation of the individual charts without affecting the other charts.

This action is identical with the behaviour of a click on the ⚓ icon in **link-lock group** of the object bar.

Lock on ....

From the menu: **Window → Lock on ....**

You can set the chart to continuously display an object in the center. Locking on an object only makes sense when you've set the coordination grid of your chart to Alt-Az together with the **"Auto-refresh every"** checkbox in the Date / Time dialog box checked. If one of these conditions is not met, your displayed chart isn't moving at all.

This action is identical with the behaviour of a click on the ⚓ icon in **link-lock group** of the object bar, or by a right click in the chart followed by a click in the pop-up window on **Lock on .....**.

Next Chart

From the menu: **Window → Next Chart**

When you 've got more charts open in your program, you can activate the next chart by a click on this line.

The short cut for this action is a combined press of the CTRL-Tab keys.

Cascade

From the menu: **Window → Cascade**

When you activate this, the charts will be positioned as a shifted stack.

This action is identical with the behaviour of a click on the ⧉ icon in **the window group A** of the main tool bar.

Tile Horizontally

From the menu: **Window → Tile Horizontally**

By activating this, the charts will be displayed from the top to the bottom.

There are no icons or other ways to get here.

Tile Vertically

From the menu: **Window → Tile Vertically**

The charts will be postioned side by side.

This action is identical with the behaviour of a click on the ☐ icon in **the window group A** of the main tool bar.

Maximize

From the menu: **Window → Maximize**

After this action the active chart will fill the entire CdC window. This is the opposite action of a **restore size**.

This action is identical with the behaviour of a click on the black square icon at the top right corner of a chart that isn't maximized

Chart List

From the menu: **Window → [Chart_#]**

A list of the opened charts appears at the bottom of the menu. Click on a corresponding line to set this chart as the "active" one.

Update Menu

This menu let you keep current some data that require to be regularly updated. You must be connected to the Internet before to use this functions.

Search for software update

Check if a new version of the Skychart software is available and propose to download.

This check only for a new stable version, except if you are already running a beta version.

Comet elements

Update the comet elements using your current setting.

This open the same dialog as Setup → Solar system → Comet but do not touch any button as all is automated here!

Asteroid elements

Update the asteroid elements using your current setting.

This open the same dialog as Setup → Solar system → Asteroid but do not touch any button as all is automated here!

Artificial satellites

Open both the https://www.space-track.org [https://www.space-track.org] web page in your internet browser, and the location where you need to download the TLE files with .tle extension in your file explorer.

This is the same function as the **Download TLE** button of the artificial satellites page of the Calendar.

Help Menu

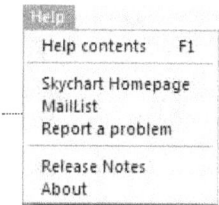

Help Contents

Here it is! Maybe you are viewing this help documentation from your local computer that came with your SkyChart installation. Keep in mind that the most recent help documentation is always available online [http://www.ap-i.net/skychart/en/documentation/start]. The online documentation is written for the most recent SkyChart version.

FAQ

A list of common problems and their solutions.

You can find the most up to date FAQs here.

Quick start guide

A basic help for beginners with SkyChart. If you need more accurate descriptions for all commands, see the Reference Manual instead.

Skychart Homepage

The official Homepage of the SkyChart [http://www.ap-i.net/skychart/start] site.

Mail List

The SkyChart Yahooo! group [http://tech.groups.yahoo.com/group/skychart-discussion/] is the place where you can exchange all sorts of information regarding SkyChart.

Report a problem

.. by using the SkyChart Mantis bug tracker [http://www.ap-i.net/mantis/view_all_bug_page.php?page_number=1].

Release Notes

A local version note. All version notes can be found in SkyChart Site **news [http://www.ap-i.net/skychart/en/news/start]** page.

About

Three tabs:

- **About** contains the version number and the date of compilation.
- **Authors** self-explanatory.
- **Licences** Text of the GNU licence.

Pop-up windows

This matrix shows the results of on-the-chart mouse actions:

action	target	opens on the chart:
Left click on a	displayed object	a label
Right click on a	displayed object	the chart pop-up window with extra object possibilities
Left click on a	label (**edit label mode: on**)	the detailed info window
Right click on a	label (**edit label mode: on**)	the editable label pop-up window
Right click on a	label (**edit label mode: off**)	the chart pop-up window
Right click on a	empty spot on the chart	the chart pop-up window

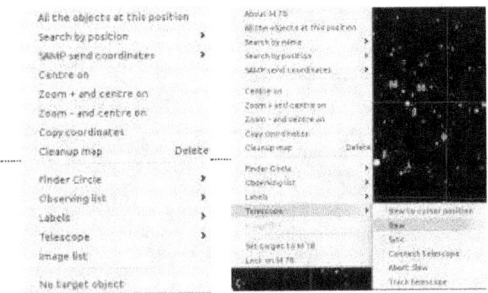

The Chart pop-up window

When the "edit label" mode is set to "off", the chart pop-up windows are shown when you do a right-click anywhere on the chart. Entries in the pop-up window are:

- **About ...** this entry only appears if you do a right-click on an object on the chart. Clicking this line opens the detailed information window containing specific information about the object.
- **All the objects at this position** This open a list of the the objects near the cursor position.
- **Search by name** Search the selected object name in external indexes.
- **Search by position** Search in external indexes for objects at the cursor position.
- **SAMP send coordinates** Send the cursor coordinates to other SAMP applications.
- **Centre on** This moves the chart to centre the position where the right click occurred.
- **Zoom + Centre** same as "Centre" and a division of the FOV by 2.
- **Zoom - Centre** same as "Centre" and a multiplication of the FOV by 2.
- **Copy coordinates** copy the current object or cursor position to the clipboard.
- **Cleanup map** Remove all temporary drawing from the map. Also remove object locking.
- **Specific tool box menu** If your current Tool box script include a menu it is visible here.
- **Finder Circle** shows a secondary menu :
  - **Select Circle** select the circle to activates.
  - **Select Rectangle** select the rectangle to activates.
  - **Eyepiece vision** simulate the vision in an eyepiece using the wider selected circle. Work only if the current chart FOV is about the same as the circle diameter.
  - **New Finder Circle** shows a floating finder circle. You can fix this to the chart by a left-click on the chart. You can create as many Finder Circles as you want. These Finder circles are independent of the "Chart → Lines/Grid → Show Mark" command or the ⊙ icon in the object tool bar.
  - **Remove Last Circle** deletes the last finder circle created by the previous command.
  - **Remove All Circles** deletes all finder circles created by the "New Finder Circle" command.
  - **Save to file** record all the circles position to a file.
  - **Load from file** read the circles position from a file.
- **Observing list** shows a secondary menu :
  - **View Observing list** Open the Observing list window.
  - **Add ... to observing list** Add the selected object to the observing list.
- **Labels** shows a secondary menu :
  - **New Label** puts a user defined label on the chart. See **Labels**.
  - **Remove Last Label** deletes the last label created by the "New Label" command.
  - **Remove All Labels** deletes all the labels created by the "New Label" command.
  - **Recover hidden labels** Recover the label hidden using the Edit Label pop-up menu.
- **Telescope** shows a secondary menu :
  - **Slew to cursor position** Move the telescope at the current position, even if no object is selected.
  - **Slew** same command as Telescope → Slew or the ⇖ icon

- **Sync** same command as Telescope → Sync or the 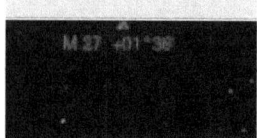 icon
- **Connect** same command as Telescope → Connect or the ⬆ icon
- **Abort Slew** Stops the execution of a "Slew" command.
- **Track telescope** same as Telescope → Track telescope.
- **Image list** If some image are displayed this open the Image list window.
- **Set target to ...** Set the current object as "target". A target indicator will be show on the screen border when the object is no more in the field.

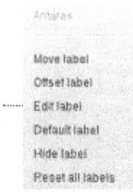

- **Lock/Unlock Chart** same as **Window** → **Lock on ....** or the ⚓ icon

The Edit Label pop-up window

Antares

Move label
Offset label
Edit label
Default label
Hide label
Reset all labels

The **Edit Label** pop-up window is only shown when you right-click a label while the "edit label" mode is set to "on". You can change this mode setting by clicking the ⍺ icon in the object tool bar, or from the menu by clicking **Edit → Edit label**. For further information, see **Labels**.

Main Bar

For many buttons on the tools bar you can access the corresponding setting by a right click on the button.

The File Group

- 🗋 **Create a new Chart** is a shortcut for **File → New chart**.
- 🗁 **Open a chart** is a shortcut for **File → Open**.
- 🖫 **Save the current chart** is a shortcut for **File → Save as...**.
- 🖨 **Print the chart** is a shortcut for **File → Print**.

Night Vision color (View)

- ☼ **Night vision color** is a shortcut for **View → Night Vision**.

The Window Group A

- ▯ **Cascade** is a shortcut for **Window → Cascade**.
- ▯▯ **Tile vertically** is a shortcut for **Window → Tile Vertically**.

The Edit Group

- ↺ **Undo Last Change** is a shortcut for **Edit → Undo**.
- ↻ **Redo Last Change** is a shortcut for **Edit → Redo**.

The Zoom Group (View)

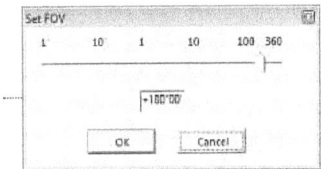

Here you can change the Field of View (FOV) from the menu bar. You can also change the FOV by turning the mouse wheel (if available). You can also zoom in by drawing a box on the chart with the mouse while pressing the left mouse key, followed by a left mouse click in the drawn area.

- +🔍 **Zoom in** is a shortcut for **View → Zoom in** (divide FOV by 2).
- -🔍 **Zoom out** is a shortcut for **View → Zoom out** (multiply FOV by 2).
- 🔍 **Set FOV** is a shortcut for **View → Set FOV**. (Set manually the FOV)

Magnitude Group

These group of buttons enable you to increase or decrease the number of objects (Stars and Deep Sky Objects) shown on your map. For every click on the Star buttons you change the limit of shown stars by a magnitude of 0.5. Also, for every click on the Deep Sky buttons you change the limit of shown Deep Sky Objects by a magnitude of 0.5.

- ▪ **More stars** is a shortcut for **Chart → Number of Stars → More Stars** (Affects Solar system objects too)
- · **Less stars** is a shortcut for **Chart → Number of Stars → Less Stars** (Affects Solar system objects too)

- ⬚ **More Deep Sky** is a shortcut for **Chart → Number of Stars → More Deep Sky**
- ⬚ **Less Deep Sky** is a shortcut for **Chart → Number of Stars → Less Deep Sky**

Search Group

[▼] 🔍 There are no other ways to get to the simple search input area.

The input area facilitates a simple search. You have to enter an object id in this catalog. A list of previously searched objects is maintained. For example, click the input area and fill in M42, hit the enter button and see what happens. Again, click the input area and enter Betelgeuse, hit enter and wonder.

🔍 **Advanced Search** is a shortcut for **Edit → Advanced Search**.

This button gives access to the Advanced Search dialog box which is detailed here.

Position

🔍 **Position** is a shortcut for **View → Position**.

Click on this icon and you can easily and quickly read or define the chart center in equatorial or azimuthal coordinates, the FOV and the rotation of the chart as described here.

Objects List

▦ **Object List** is a shortcut for **View → Object list**.

Click on this icon to obtain a catalog excerpt of all objects displayed on the chart. See **Object list** for details.

Observing List

▭ **Observing List** is a shortcut for **File → Observing list**.

Click on this icon to open the Observing List window.

Calendar

📅 **Ephemeris Calendar** is a shortcut for **File → Calendar**.

Solar System Information

📑 **Solar System Information** is a shortcut for **View → Solar System Information**.

Time Simulation group

◁ ☐ ▷ ▷| [1 ▲▼] [Hour ▲▼] A faster way to modify the date and time than **Setup → Date / Time**.

With the two comboboxes you can choose the step unit and increment. By clicking the arrows you can go forward or reverse relative to the displayed time. A click on the square icon between arrows sets the time to "now" (system date/time).

The ▷ button start or stop the animation with the given time step. Right click on this button to access the animation and recording setup.

Related to:

Chart Coordinate System When you want to see what is visible above your horizon at a certain time and want to see time simulation in effect, the Chart Coordination System needs to be set to Azimutal Coordinate. There are several ways to achieve this. For example, from the menu bar: **Chart → Chart Coordinate System → Alt-Az Coordinate**

Solar System Probably you want to simulate movement of planets, asteroids or comets in whatever coordinate system. You need to set them visible, also here there are several ways. From the menu bar: **Chart → Show Objects** and check "Show Planets", "Show Asteroids" and "Show Comets. Or switch the display status for those objects by clicking the proper icon on the objects bar.

Telescope Group

- 🜚 **Contol Panel** is a shortcut for **Telescope** → <u>Control Panel</u>
- ⊕ **Sync** is a shortcut for **Telescope** → <u>Sync</u>
- ⛭ **Slew** is a shortcut for **Telescope** → <u>Slew</u>
- 🜨 **Abort Slew** is a shortcut for **Telescope** → <u>Abort Slew</u> or the keys Ctrl+K

Window Group B

- ⊡ Restores the active chart to the previous layout. The oposite action of a <u>**maximize**</u>.
- **x** **Close** is a shortcut for **File** → <u>Close chart</u>. Close the active chart (only if there's more than one chart open).

Object Bar

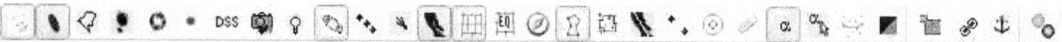

The object bar is displayed horizontally, just above the chart, under the main bar. It contains the icons that are related to the display of all kinds of things on the chart.

For many buttons on the tools bar you can access the corresponding setting by a right click on the button.

Object Group A

- **Show stars** is a shortcut for **Chart → Show objects → Show stars**.
- **Show Deep Sky Objects** is a shortcut for **Chart → Show objects → Show Deep Sky Objects**.
- **Show Lines** is a shortcut for **Chart → Show objects → Show Lines**.
- **Show Pictures** is a shortcut for **Chart → Show objects → Show Pictures**.
- **Show VO data** activate the display of the Virtual Observatory catalog
- **Show user objects** activate the display of the User defined objects

Pictures Group

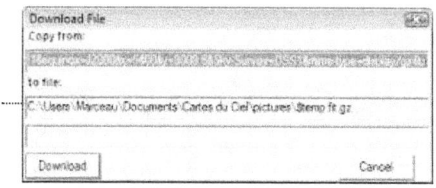

- DSS **Get DSS Image** is a shortcut for **Chart → Get DSS Image**.
- **Change pictures display** is a shortcut for **Setup → Pictures**.
- **Blinking Picture** is a shortcut for **View → Blinking Picture**.

Object Group B

- **Show Planets** is a shortcut for **Chart → Show objects → Show Planets**.
- **Show Asteroids** is a shortcut for **Chart → Show objects → Show Asteroids**.
- **Show Comets** is a shortcut for **Chart → Show objects → Show Comets**.
- **Show Milky Way** is a shortcut for **Chart → Show objects → Show Milky Way**.

Lines - Grid Group

- **Show Coordinate Grid** is a shortcut for **Chart → Lines/Grid → Show Coordinate Grid**.
- **Add Equatorial Grid** is a shortcut for **Chart → Lines/Grid → Add Equatorial Grid**.
- **Show compass**
- **Show Constellation Line** is a shortcut for **Chart → Lines/Grid → Show Constellation Line**.
- **Show Constellation Limit** is a shortcut for **Chart → Lines/Grid → Show Constellation Limit**.
- **Show Galactic Equator** is a shortcut for **Chart → Lines/Grid → Show Galactic Equator**.
- **Show Ecliptic** is a shortcut for **Chart → Lines/Grid → Show Ecliptic**.

Marks Group

- **Show Mark** is a shortcut for **Chart → Lines/Grid → Show Mark**.

- 🖊 **Distance measurement** button:

When "Distance measurement" is down, the mouse left click behavior change.
Click on the first point, without releasing the button move the cursor to the second point and release the button.
The distance, position angle, RA and DEC offset is displayed on the left of the status bar.
The first and second point can be any position even without an object. But if an object is identified near the cursor its center position is used instead of the mouse cursor position. In this last case the identification label is show near the object.
Click the "Distance measurement" button again to return to normal cursor use.

- α **Show Labels** is a shortcut for **Chart → <u>Show Label</u>**.
- α **Edit Label** is a shortcut for **Edit → <u>Edit Label</u>**.
- 🌙 **Show Object below the horizon** is a shortcut for **Chart → <u>Below the horizon</u>**.
- ◼ **Sky Background Color** is a shortcut for **View → <u>Sky Background Color</u>**.

Link-Lock Group

- 🎞/✣ **Change mouse mode** Change the left mouse click behavior: Zoom windows selection, or move/pan.
- 🔗 **Link all chart** is a shortcut for **Window → <u>Link All Chart</u>**.
- ⚓ **Lock on ...** is a shortcut for **Window → <u>Lock on...</u>**.

Drawing mode

- ✎ **Change Drawing mode** this tool switches the drawing mode of stars and deep sky objects between the three modes: "line", "photographic" and "parametric". See **Setup → <u>Display Mode</u>**.

The bar at the left

Observatory

- 🔧 **Set observatory location** is a shortcut for **Setup → Observatory**.

Set date and time

- 🕐 **Set date and time** is a shortcut for **Setup → Date / Time**.

Setup

- 🗔 **Configure the program** is similar way to get to **Setup**.

Coordinate System group

- EQ **Equatorial coordinate** is a shortcut for **Chart → Chart coordinate system → Equatorial coordinate**.
- AZ **Alt-Az coordinate** is a shortcut for **Chart → Chart coordinate system → Alt-Az coordinate**.
- EC **Ecliptic coordinate** is a shortcut for **Chart → Chart coordinate system → Ecliptic coordinate**.
- GI **Galactic coordinate** is a shortcut for **Chart → Chart coordinate system → Galactic coordinate**.

Equinox and epoch settings can be done by **Setup → Chart, Coordinates**

Transformation Group

- ◁▷ **Miror horizontally** is a shortcut for **Chart → Transformation → Miror horizontally** (the icon changes to red if the chart is mirrored).
- ⇕ **Miror vertically** is a shortcut for **Chart → Transformation → Miror vertically** (the icon changes to red if the chart is mirrored).
- ↻ **Rotate right** rotates the chart 15° clockwise, is a shortcut for **Chart → Transformation → Rotate right**. See also Main bar → **Position**.
- ↺ **Rotate left** rotates the chart 15° counter-clockwise, is a shortcut for **Chart → Transformation → Rotate left**. See also Main bar → **Position**.
- ↕ **Rotate by 180°** is a shortcut for **Chart → Transformation → Rotate by 180°** This is how you can get mirror horizontally+vertically. (the icon changes to red if the chart is rotated).
- Right click on any of this button to **reset the orientation**.

The bar at the right

The Field of Vision group

These icons are shortcuts for **Chart** → **Field of Vision** → ....

Here you can set the width/height of the displayed field of vision of your active chart to a predefined value (in degrees of arc). A click on the ⊙ icon sets the FOV to 360°.

You can change the default settings for the fields of Vision by **Setup** → **Chart, coordinates** → **Field of Vision**.

The horizon group

N
S
E
W
Z

These icons are shortcuts for **Chart** → **View Horizon** → ...., with the exception for the Zenith icon.

If you are interested in the objects display on the chart situated in one of the cardinal directions, simply click the appropriate icon to make your choice:

- **N** for North,
- **S** for South,
- **W** for West,
- **E** for East,
- **Z** for Zenith

Setting the chart by these methods to one of the cardinal directions, also forces the chart to use the Altitude-Azimuth coordinate system.

Tool box

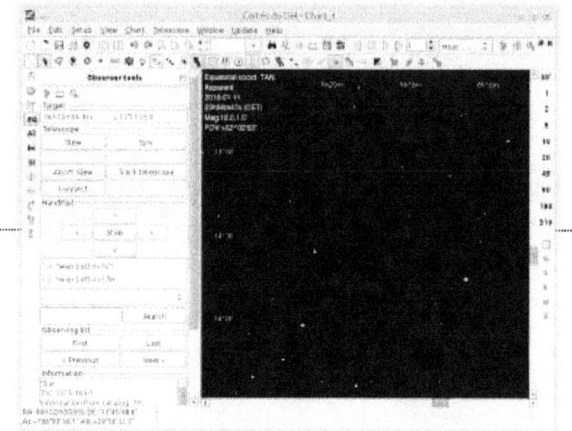

The Tool box contain functions for specific task you can select depending on your needs. You can also add a new function by using the script editor.

There is up to eight tool box you access from the main menu View → Tool box, or using the keyboard shortcut F1 to F8. Press the same key again to hide the left panel.

The Tool box are show vertically on the left of the screen, this let enough space for the chart itself with most of the wide screen display. The Tool box width can be adjusted with the mouse, and if your screen is too small a scroll bar can be used.

Each Tool box contain three part:
- Up to two button bar where you can add the same button as in the tool bar.
- The main area with specific button and text box.
- At the bottom, two buttons allow to select or edit the scripts.

Open the menu **Setup→Manage toolbox** or click the Manage toolbox button.

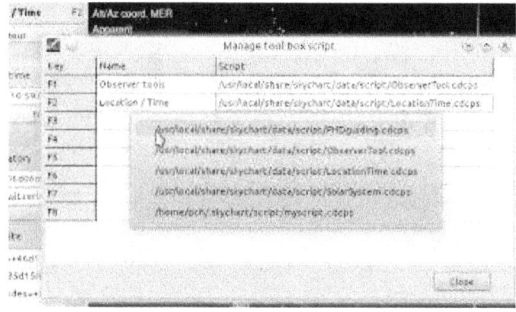

Click on the Script column to select the tool you want for the corresponding F key.
Right click on a row to open a menu that allow to open a script at another location or to remove a script assignment.

You can use the tool box and script editor to create your own script or to make a modification to one of the tool.

Standard tools

Observer tools

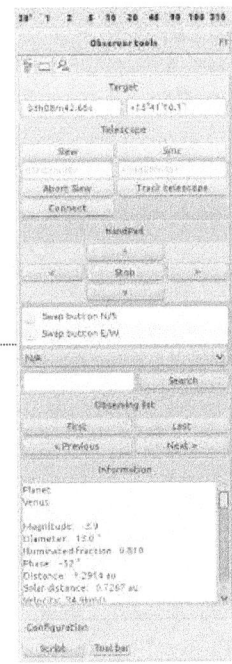

This panel contain convenient functions when you use the program connected to a goto telescope.

It contain the telescope control function, including a virtual handpad.
To use the handpad select first if the direction button need to be swapped and select the speed (in degree/second) from the list. Then click one of the direction button to start the move, and click the Stop button to stop.

You can navigate your observing list or search for an object name.

Each time an object is selected the "Target" coordinates on the top are updated and ready to use with the telescope buttons.

The Information box give the detailed information about the object without the need to open a separate window.

Solar System

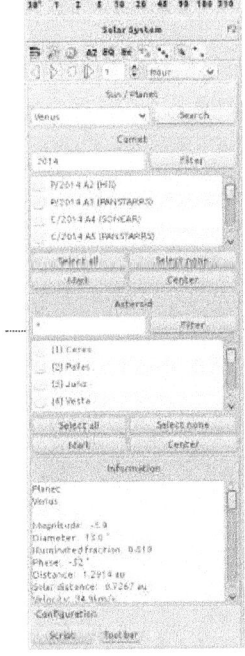

This panel group the information about the Solar System objects.

The tool bar contain the related buttons so you not need to add them to the main screen.

You can select a planet, comet or asteroid from the specific box.

For the comet and asteroid you can filter the list using your criteria.
Click on a row and the Center button to center the object on the map.

You can also select some objects in the list and click the Mark button to draw a circle on the map at the object location, independently of the object visibility .

The Information box give the detailed information about the object without the need to open a separate window.

Location / Time

This panel contain the tools to quickly change the date and the observer location.

The tool bar contain the related buttons so you not need to add them to the main screen.

The first group allow to change the date and time. Use the negative sign format for BC years.

The next group contain the information about the current location.
A drop down list allow to quickly change for an UTC time zone.

Then your favorite location are listed for selection.
You can also load an other list from a file.
A few interesting antic location are added at the bottom of the list to serve as an example for the file format.

The Information box give the detailed information about the selected object without the need to open a separate window.

PHD Guiding

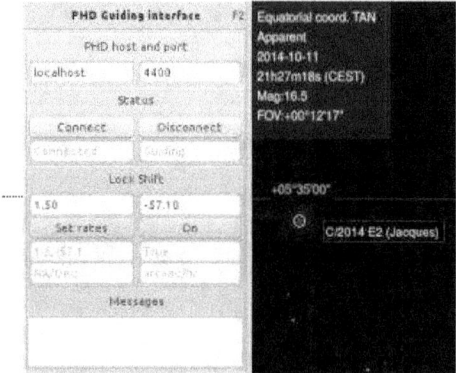

This panel contain the tools to interface with PHD2 Guiding [http://openphdguiding.org/].

The interface let you easily use the ""lock shift" function to keep your telescope pointed on a moving object while guiding on a star.

You can also activate the "dithering" function to apply a random offset to your guide position.

Connection to PHD

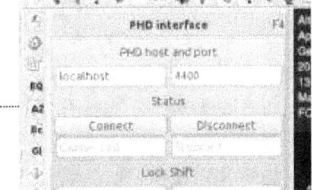

Start PHD2 and check that Tools / Enable server is checked.

In Skychart you can change the host on which PHD is running and the connexion port. If Skychart and PHD run on the same computer you can just keep the default values.

Click the "Connect" button. The status must change to "Connected" and indicate the current state of PHD.

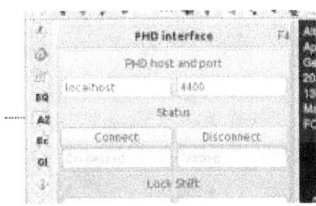

Initialize PHD as usual and start guiding. The state must change to "Guiding".

Use the Lock shift function

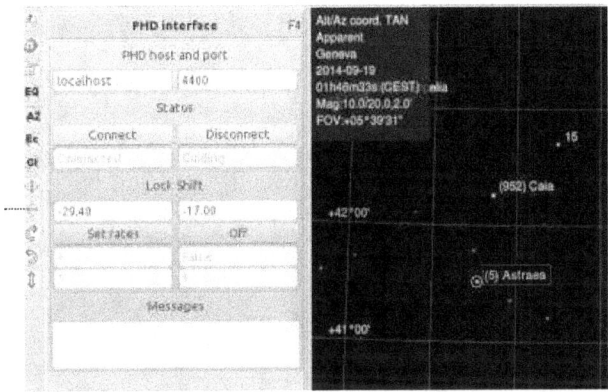

This function allow you to keep your telescope pointed on a moving object while guiding on a star.

Click on a moving object in the map, for example an asteroid or comet. The object displacement in arcsec/hour is show.

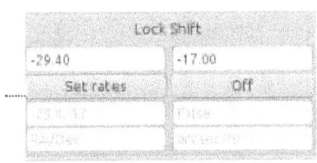

To send this displacement rate to PHD click the "Set rate" button. The current PHD lock shift status is displayed to confirm the change.

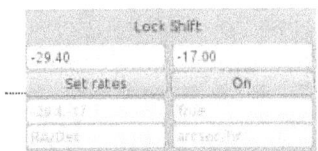

To start the shifted guiding on the moving object click the "Off" button. The button text change to "On" and the PHD status change to "True".

Use the Dithering function

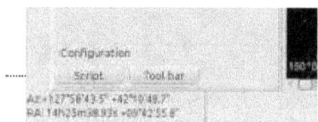

The dithering function let you apply a random offset to the guiding position to avoid to cumulate the sensor defects at the same position on every exposure.

The function is disabled by default because this is normally done by the capture software between the exposure. The use of this function from Skychart is only interesting if you take your pictures with a standalone camera or with a software that not support dithering.

To activate the function, open the script configuration with the "Script" button.

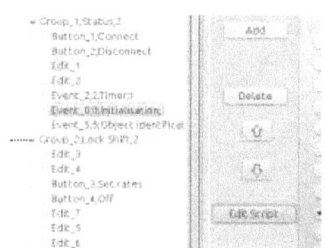

Select the row "Event_0:0:Initialisation" and click the "Edit script" button.

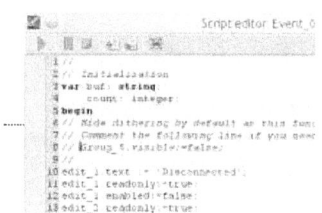

At the top of the script comment out the line that hide the dithering box. Just add // at the beginning of the line.

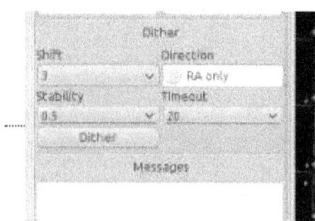

Save the modified script and click the Apply button. Save the Skychart configuration for a permanent change.

To apply an offset set the required parameters value and click the Dither button. **Be sure you click this button only between exposure when the camera shutter is closed!**

Add other functions interface

You can use the provided scripts as an example to interface with other function provided by PHD2.

The PHD wiki [https://code.google.com/p/open-phd-guiding/wiki/EventMonitoring] give all the necessary informations about this functions.

Tool bar editor

You access this function from the main menu Setup → Tool bar editor or with a right mouse click on an empty area of a tool bar.

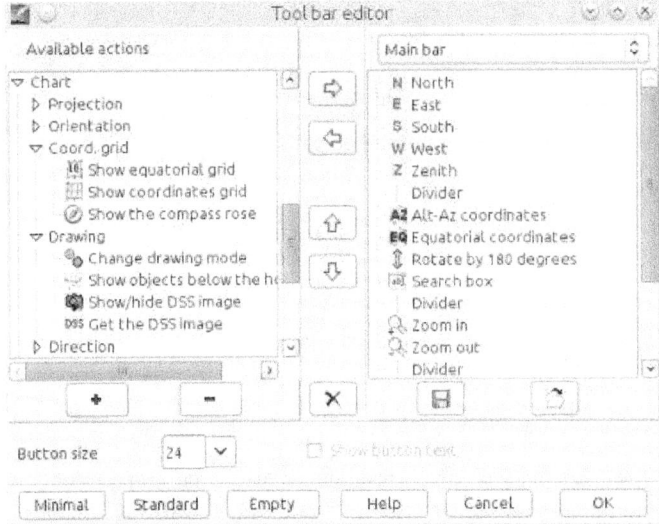

The available function are listed on the left, the currently configured tool bar are on the right. The functions on the left are grouped as in the menu.

Use the selection box on the top right to select the bar you want to edit.

- To add a new button, select the function you want to add on the left, the position of the new button on the right, then click the right arrow button.
- To remove a button, select the button on the right, click the left arrow button.
- To change the position of a button, select the button on the right, click the up or down arrow button.
- To move a button to another bar you need to remove it from the first bar and then add to the second.
- The "Quick search" and the "Time Increment" selection are not intended to be placed in a vertical bar.

You can expand the left list with a click on the "+" button, or collapse it with the "-" button.

There is three buttons to set a preconfigured layout:

- **Minimal** A single row of button intended for the observation at the telescope.
- **Standard** The same button configuration as in the previous program version. Use this button to return to the default setting.
- **Empty** Clear all the bar to allow you to start your own layout.

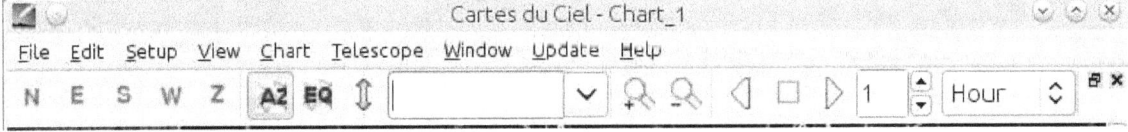

You can also increase the size of the button to make them more easy to click on the high resolution screen.
If you not add too much button for your screen size and if the button size is at least 40 you can also display the related text to help to learn the program functions.

When ready click the **OK** button to apply your change. Use the menu Setup → Save configuration now, to save your setting for a future session.

You can always select to show or hide a full tool bar from the menu View → Tool Bar, but an empty bar will never be show.

Status Bar

You will find the status bar at the bottom of your SkyChart window. It is split in two: the coordinates part at the left, an information summary at the right hand side.

Left

Az:+49°05'12.8" -05°11'42.6"
AR: 20h14m21.29s +22°39'18.2" ..... Here you can find two lines which display the coordinates of the current cursor position at the

chart. The line on top will be in the coordination system as you set the chart. The line at the bottom will display the position according to the equatorial coordination system, except when you set your chart to use the equatorial coordination system. Then it will display the position in the altitude - azimuth coordination system.

You can easily change the coordinate system of the displayed chart with the buttons of the **coordinate system group** at the left of your chart.
For the displayed equatorial coordinates in your chart, you can also choose the **equinox**
[http://en.wikipedia.org/wiki/Equinox_(celestial_coordinates)] to base your equatorial coordinate system on by the tab **Chart, Coordinates** from the **Setup → Chart, Coordinates** dialog box.

Right

After you selected an object, this part of the status bar will display a summary of the detailed information from the catalog about this object:

RA: 04h36m45.23s DE:+16°32'05.2" Star: Alp Tau Visual magnitude: 0.87 Common name:Aldebaran HD:29139
Rise: 12h47m Culmination: 20h01m Set: 3h20m

- Equatorial coordinates of the object. You can determine which of the equatorial coordinate types is to be displayed by setting this in the lower part of the first tab in **Setup → Chart, Coordinates → Chart, Coordinates**.
- The first part of the **detailed information** (It will start with the catalog short name, the object identification and further characteristics of the object. The type is coded as defined in the list of objects.)

Detailed Information

How to retrieve detailed information

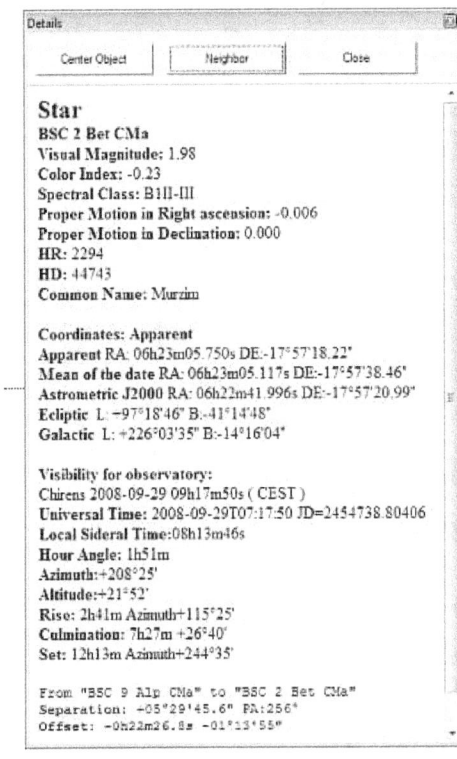

- Click on a line of the **Object List** to get detailed information about the object.
- Right-click on an object at the chart and select "About …" in the **pop-up window**.
- When *edit label* mode is **on** (the icon in the **object toolbar**): Click on an object label.
- Get a "Neighbour" window by a click on the **neighbour** button in an information detail window and click on a line as with an **Object List**.

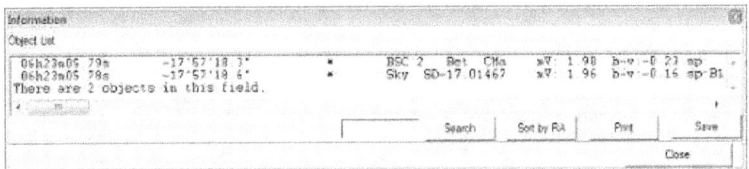

Content of the detailed information window

In any detailed information window you can distinguish 4 paragraphs with their specific sorts of data.

Object information and characteristics

For stars and deep sky objects, this paragraph contains static information about the object which was retrieved from the catalog. For these kinds of objects, they ususally will begin with the object type, catalog short name, the catalog identification, and the relative magnitude. The rest of this content depends entirely on the catalog from which the data is retrieved.

For Solar system objects, you can find information like the object type, and the object identification (name). Please forgive SkyChart when it shows you the Sun as a planet. All other information in this paragraph is dynamically calculated from the ephemeris data. This information depends on the type of solar system object you 're dealing with.

Coordinates

The first line of this paragraph, starting with `Coordinates:`, informs you to which of the equatorial coordinate systems the chart is set. The chart can show you on several coordinates in the equatorial systems, for example at the **status bar**. You can specify the equatorial coordinate system of your chart by the tab Chart, Coordinates from the menu by **setup → Chart, Coordinates**, in the **type of coordinates** part.

The other lines show you the coordinates of your object in the specified coordinate systems. Here, you can find the equatorial systems (apparent and mean), and also in the ecliptic - and galactic systems.

Visibility for your observatory position and time

Here you find specific data relating to your position and time. So there is the name of the site you are observing from with the **local date** and **time**. Next, you will find the times in **UT** and **local sideral time**. Then there is the positional data for the date and time that was used in the calculation: **local hour angle**, **azimuth** and **altitude**. At the and you can find data about the time and position of **rise**, **culmination** and **set**; unless you are dealing with a circumpolair object. Then there's only information about the **culmination**.

Distance and angle

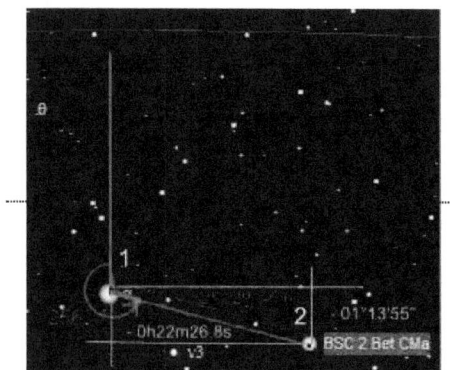
Here you can find the distance and angle between two objects that you consecutively selected. This data can only reliably calculated when you select the objects from one single calculated chart or list. (A time difference combined with the dynamics of solar system objects can't lead to a precise value.) So, in order to retrieve distances, it can be wise to uncheck the **Auto-refresh every** checkbox, or to set the auto refresh time to a value so that there's time enough for you to select your objects.

First there is the line that says **which objects** you selected for the measurement. The next line shows the **seperation** in degrees of arc in sexagesimal format, and the **angle** measured from the previous to the last selected object. This angle takes the celestial North as zero reference, and increases by East, South, West. The picture is pretty self explanatory. About the angle: when you are on the Northern hemisphere and your face is directed south, the East will be at the left hand side. The last line shows you the **offset** in sexagesimal hours and sexagesimal degrees in the equatorial coordination system.

Object List

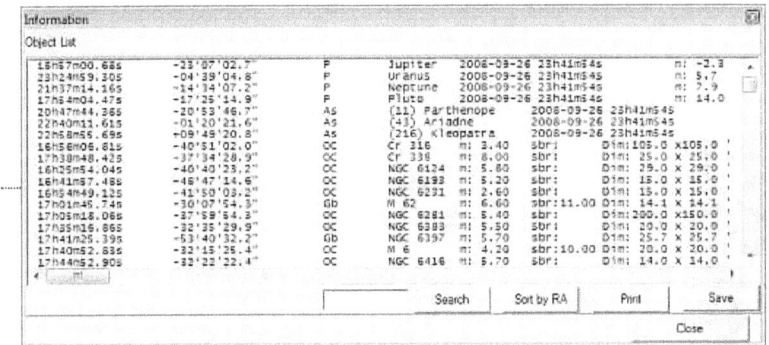

Click on the ⊞ icon from the **main bar** to retrieve an exerpt from the catalogs for the displayed objects on the chart. To configure which object types are to be listed, you need to adapt the settings to your needs by **Setup → Chart, Coordinates → the Object list Setting** tab.

Every row in this list corresponds to one of the displayed objects on the chart. At the bottom of the list the total amount of displayed objects is listed.

If you click on one row, a window is displayed which contains the **detailed** and **labeled** information about this particular object. This is the same information that you would receive from a **pop-up window** caused by a right click on the object in the chart, followed by a left click on the "About ..." entry in the the pop-up menu.

This list can be printed and saved as a .CSV file. You can search these lists for any particular object by using the input area or sort the list by RA.

Solar system information

This window give information about the planets, their visibility and their orbits.

Planet visibility

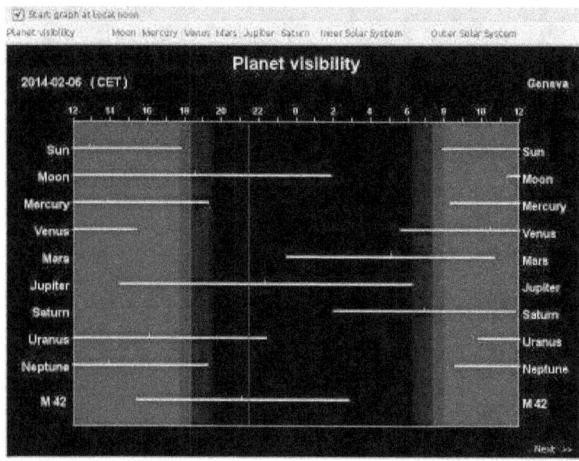

Show a graph with the planet visibility for the current day. The blue band mark the twilight time and for each object the yellow bar indicate it is visible above the horizon.
You can select to start the graph at noon for a better overview of the night time.
The visibility chart also include the currently selected object, even for stars or DSO.

Planet view

Show the current aspect of every planets.

Inner solar system

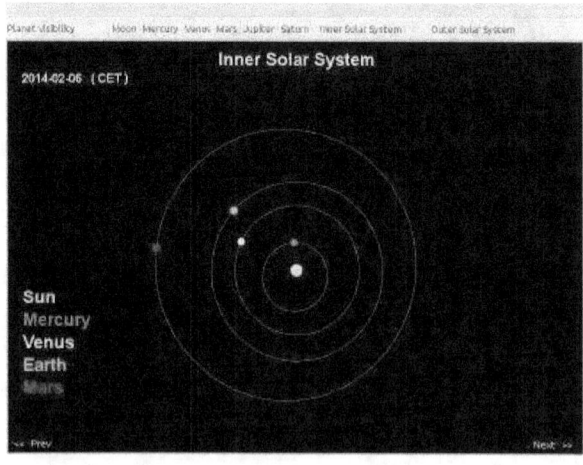

Show the orbit and position of the planets from Mercury to Mars.

Outer solar system

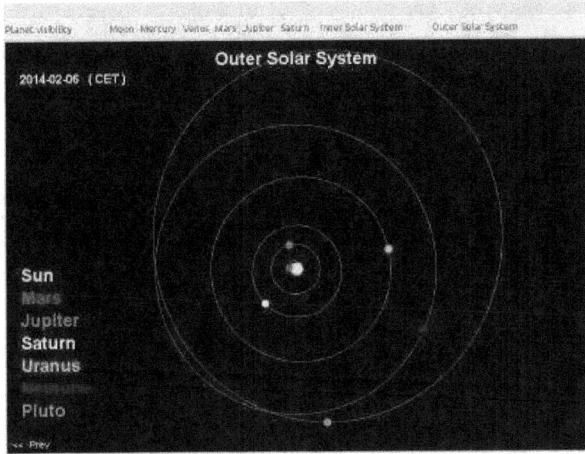

Show the orbit and position of the planets from Mars to Pluto.

Calendar, Input Area

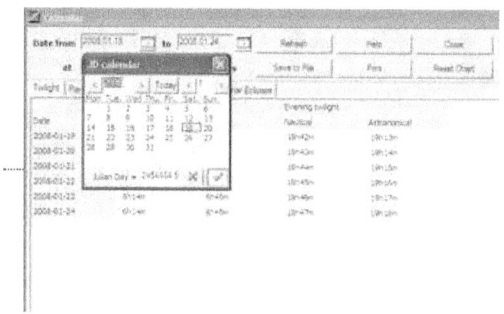

You can select the start date, end date, and the number of days between each calculation, and the time when the positions are calculated. By default, the calculations are from the current date until the next five days at 0h UT. Beware of the calculation time, when you select a long timespan.

You can click any element of the table to display the corresponding chart. The program shows a chart using the selected date and time, and centers the selected object. If the selected column contains the time for a specific event like a planet rise time, or the twilight time, this time is used for the chart. When you change the used time this way, you will receive a warning about this. Entering "Yes" confirms to use the new time. When you enter "No", the list will not close to give you the opportunity to push the "Reset" button (top-right).

Calendar, Twilight

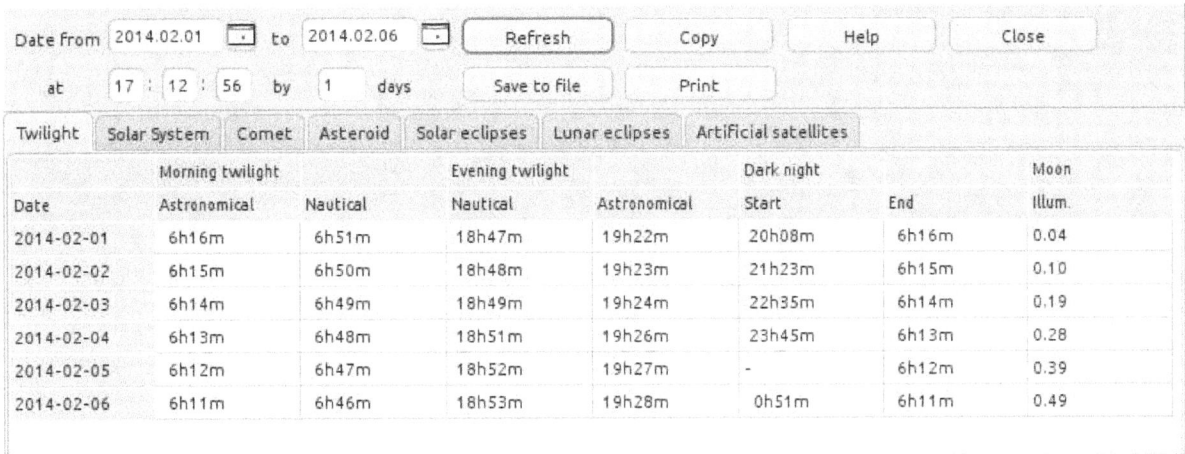

This screen shows you the morning and evening twilight [http://en.wikipedia.org/wiki/Twilight] times for the dates as you had set in the **input areas** of the calendar. Twilight is distinguished in 3 sorts:

- **1. civilian twilight** the center of the sun is between 0º and 6º of arc below the horizon. (This period of time is not shown in SkyChart).
- **2. Nautical twilight** the center of the sun is between 6º and 12º of arc below the horizon, the first stars of magnitude 2 are visible.
- **3. Astronomical twilight** the center of the sun is between 12º and 18º of arc below the horizon, the sky is fully dark for astronomical observation.

When you set your chart to use a **Alt-Az coordinate system** together with the usage of a **automatic background colour**, the sky background will display a lighter colour, related to the sort of twilight.
Astronomical dusk is the period of time during which the center of the sun is lower than 18º degrees of arc below the horizon for your observation position. For those periods, SkyChart will usually display a dark background, except when the Moon is above the horizon.

The last columns of the table show the begin and end of this dark period including for the Moon presence. The Moon illumination fraction is also indicated for a better estimation of how bad it will be.

Calendar, Solar System

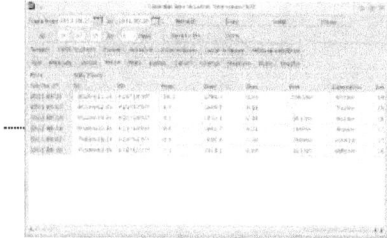

Shows the position, magnitude, apparent diameter, illuminated fraction, rise, culmination and set time. Also the current position in altitude and azimuth is shown for the planets (Pluto included), the Moon and the Sun, for each selected date.

The **graphs** tab shows the evolution in the given period of time of the rise, culmination and set times, magnitude, diameter and illumination of the planets.

Calendar, Comet

Shows the position, magnitude, Sun elongation, phase, rise, culmination and set time for the selected comets.

The elevation above horizon and azimuth at the twilight time are indicated to help observing comets near the Sun. This indicates if the comet is best observed on the morning or evening.

Before you can display or search comets with SkyChart, you must have previously downloaded a file with the orbital elements of comets. Click **here** to read more.

You can filter the comet list by name or click the "Brightest" button to sort the list with the brightest first.

Calendar, Asteroid

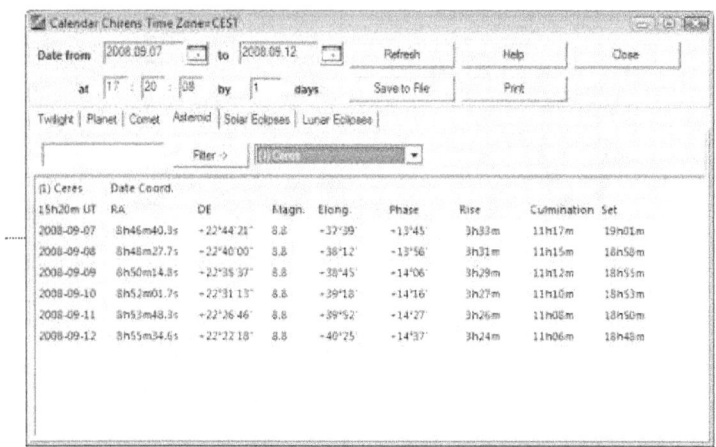

Shows the position, magnitude, Sun elongation, phase, rise, culmination and set time for the selected asteroid.
You can simply select an asteroid from a picklist by clicking the combobox. This way, you can only select from the first 500 asteroids. The good part is that you can search for any listed asteroid, simply enter (a part of) the name or number, press the filter button, and select the asteroid from the combobox.

Before you can show or search for asteroids, you must have previously downloaded a file with the orbital elements of asteroids. To learn about that, read **this**.

Calendar, Solar Eclipses

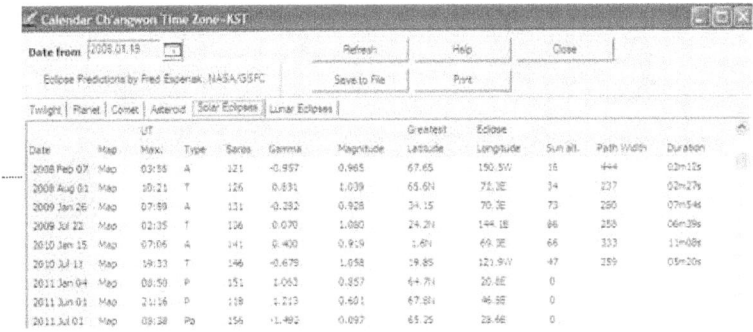

All the information in this page comes from **Fred Espenak, NASA/GSFC [http://eclipse.gsfc.nasa.gov/eclipse.html]**. You can visit his web site by clicking the large button at the top of the dialog box. The list shows the Solar eclipses for the current century. It shows the date and time of the maxima, the availability of the detailed map, the type of eclipse, the saros number, the gamma value, the eclipse magnitude, the location of the eclipse maxima, the altitude of the Sun at this point, the path width in kilometers, and the duration of the total or annular phase. See a detailed explanation of these values. The data from 1800 to 2100 are included, you can download additional data from the web page.

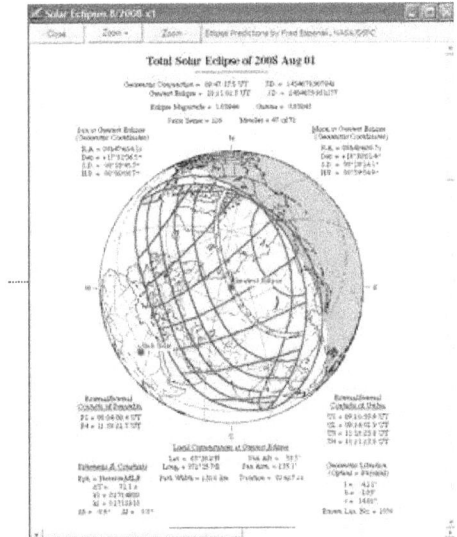

A mouse click on the list shows the eclipse from the current observatory location. If you click on the Latitude or Longitude column, this shows the eclipse from the maxima location on the World. If you click the Map column this shows the detailed map for this eclipse if available.

When you clicked a date from the list, you also changed SkyChart to use this date. When you close the Calendar dialog box, you will receive a warning about the changed time. You can click **Yes** to continue with the changed time, or click the **NO** button followed by a click on the **Reset Chart** button in the dialog box.

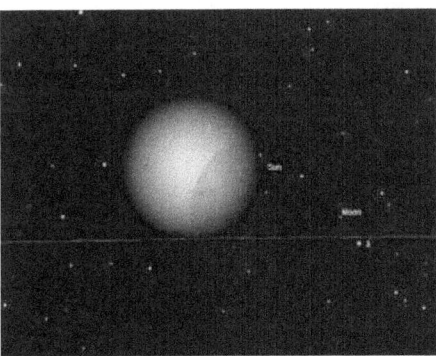

Calendar, Lunar Eclipses

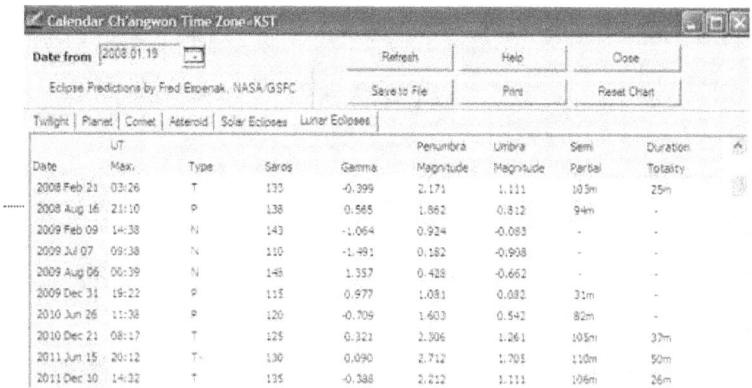

All the information in this page comes from **Fred Espenak, NASA/GSFC [http://eclipse.gsfc.nasa.gov/eclipse.html]**. You can visit his web site by clicking the large button at the top of the dialog box.

The list shows the Lunar eclipses for the current century. This shows the date and time of the maxima, the type of eclipse, the saros number, the gamma value, the magnitude of the penumbra and umbra, the duration of the partial and total phases.

The data from 1800 to 2100 are include with the basic version, you can download additional data from the web page.

A mouse click on the list shows the eclipse from the current observatory location as in `Solar Eclipses`.

When you clicked a date from the list, you also changed SkyCharts to use this date. When you close the Calendar dialog box, you will receive a warning about the changed time. You can click **Yes** to continue with the changed time, or click the **NO** button followed by a click on the **Reset Chart** button in the dialog box.

Calendar, Artificial satellites

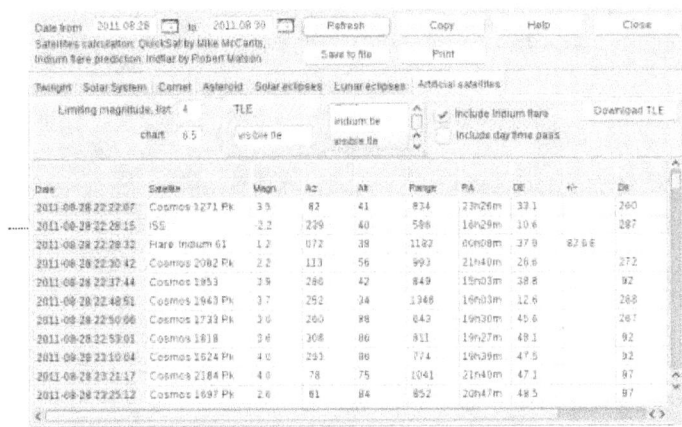

The calculation for artificial satellites uses Mike McCant's QuickSat [http://www.prismnet.com/~mmccants/] program and Robert Matson's Iridflar program. Take the time to read the documentation for these programs. It is located in the data/quicksat and data/iridflar folder.

First select the date range for the prediction, the limiting magnitude for the list and to plot to the chart, the TLE file to use for the calculation (use the Ctrl or Shift key to select multiple files). Use the two check boxes on the right if you want to see daytime pass or if you want to include the Iridium flare to the list.

The TLE file for the Iridium flare must be named iridium.tle. If you want to show the satellite path before and after the flare on the chart you must include this file on the list and select a limiting magnitude of at least 8 for the chart.

After clicking Refresh, the list shows the following values:

- the date and time of maximum elevation
- the satellite name
- the maximum magnitude
- azimuth and altitude of the maxima
- distance in kilometers
- right ascension and declination
- estimated error on the prediction time in minutes or for the flare the distance and direction to the flare maximum, and the direction of the satellite movement.

If the magnitude is between brackets this indicates that the absolute magnitude of the satellite is not known and the calculated value assumes an absolute magnitude of 6. The magnitude file "qs.mag" is updated from http://www.prismnet.com/~mmccants/ [http://www.prismnet.com/~mmccants/] along with the tle. You can also replace the file in the User settings\satellite folder manually.

Clicking on the list shows the chart of this satellite pass but also any other object brighter than the limiting magnitude in a time range of +/- 6 minutes.

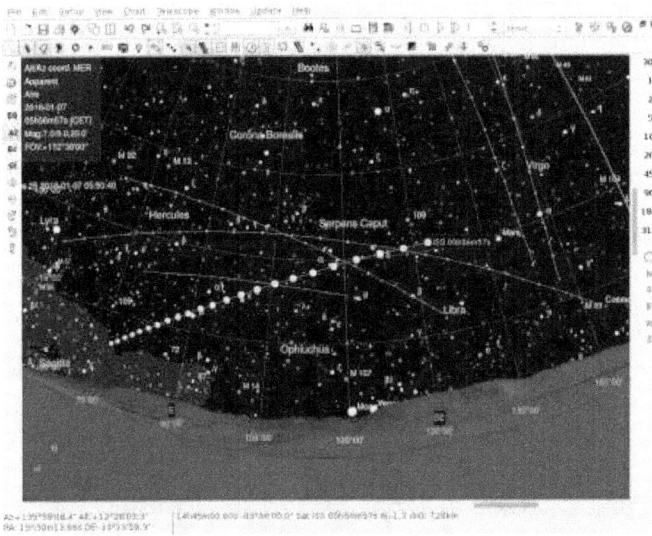

Please update the TLE element files regularly (at least every week) using the "TLE download" button.
You can select the URL source you want to download from Setup→Internet

The button "Download TLE" download and install the selected files, or if the list is empty, it show you a message with the path where you need to put the files and then open your web browser to the space-track.org web page.

You can also use a script to download the files. The following example can be used as a starting point to get the files you need. You can get tledownload for Windows [http://www.ap-i.net/pub/skychart/satellites/tledownload.zip] or for Linux or Mac [http://www.ap-i.net/pub/skychart/satellites/tledownload.tgz]. It require you create a space-track [https://www.space-track.org] account first.

Require software installation

The Quicksat program is a win32 application, it run natively on Windows 32 and 64 bits.
On Linux and Mac you need to install Wine [http://www.winehq.org/]:
Linux:

```
sudo apt-get install wine
yum install wine
```

Mac: See the specific page about Wine installation

The Iridflar program is a dos16 application, it run natively only on Windows 32 bits.
On Windows 64 bits a version of Dosbox is include with the Skychart installer.
On Linux and Mac you need to install Dosbox [http://www.dosbox.com/]:
Linux:

```
sudo apt-get install dosbox
yum install dosbox
```

Mac: See just after the Wine installation

Date / Time setting

From the menu: **Setup → Day / Time**.
You also can retrieve this dialog box by a click at the 🜨 icon from the **left bar**, or by **Setup → All configuration options → Date / Time**.
You can also access directly the Animation setting by a right click on the ▷ button.

The Date/Time setting has three tabs:

- Time
- Simulation
- Animation

For a proper display of your chart, also make sure you set the right **observation** position.

Time

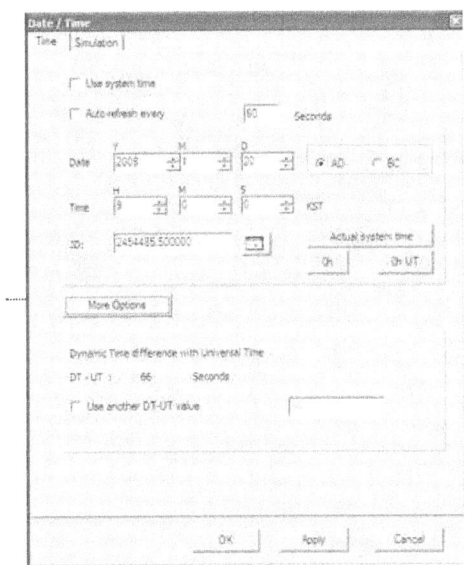

With this dialog box you can set the date and time for SkyChart to use. To display a reliable chart, SkyChart needs to do many calculations which require time as a parameter. The proper date and time setting is very important to display an accurate position of the planets, comets and asteroids. The date and time is also very important to display a proper chart when you are using Alt-Az coordination grid, or when you want to check the visibility of an object in the detailed information window. You might even want to display the proper motion of stars over many years.

The checkbox Use system time

By checking this box, SkyChart will copy the system time from your computer at the moment you check this box, or every time you open a new chart and use this time to calculate the chart. A very convenient choice when you want to display a chart of the sky at the present time.

Setting the date and time manually

You need to **uncheck** the *Use system time* checkbox when you want to display a chart **for a date / time in the future or past**. You can set the date and time by usage of the appropriate combo boxes. SkyChart can calculate with dates in the range from -20000 to +20000 years. However, the planet positions are only properly calculated for dates from -3000 to +3000. Beware with the negative years: 1BC is the year 0 and 2BC is the year -1. Note that the entered time is the local time (legal time) of the defined observatory location.
You also can set **JD time [http://en.wikipedia.org/wiki/Julian_day]** by using the input area and the JD calendar dialog box.

fixed time buttons

Tonight: Set the time to the beginning of the astronomical twilight for today.
Actual system time: At the moment you click this button, the system time will be copied and used as if you set it manually. The active chart will be recalculated with this time. From now on, every newly opened chart also will use this time.
00h: At the moment you click this button, the system time will be set to 00:00:00 of your local time. The date will be untouched. The active chart will be recalculated with this time. From now on, every newly opened chart also will use this time.
00h UT: At the moment you click this button, the system time will be set to 00:00:00 Universal Time. The date will be

untouched. The active chart will be recalculated with this time. From now on, every newly opened chart also will use this time.

The checkbox Auto Refresh

If you want SkyChart to automatically display an updated chart, you need to check this checkbox. Also, set a value in seconds for the interval after which you want SkyChart to generate a new chart. Keep in mind that generating charts can be a demanding task for your processor. Usually it only makes sense to *auto refresh* your charts when you are using the Alt-Az coordination grid in the chart. If you are observing occultations, asteroids or comets, it also can be usefull even if your chart isn't using Alt-Az grid.

By a click at the box **More Options** you get the possibility to change the difference between the Dynamic Time (DT) and the Universal Time (UT) used by the program. You need to set the correct value when you want to retrieve the accurate data regarding occultations. Click **here [http://en.wikipedia.org/wiki/Time_standard#Time_standards_for_planetary_motion_calculations]** to read more about time standards.

Simulation

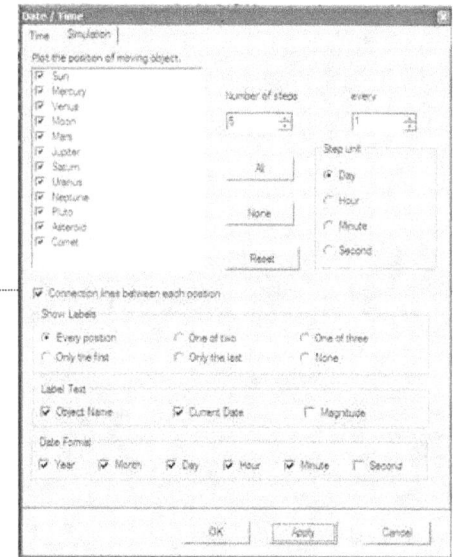

SkyChart can display the movement of the solar system objects (Sun, planets, moons, asteroids and comets) for a period of time in one single chart. The part of the orbit that those objects move during this time, can be displayed by a line and positions at a choosen interval.

You define the total period of time by setting values in the combo boxes **Number of steps**, **every** (which can be read as the **step size** or interval) and selecting a **unit** with the radio button group. If you set *Number of steps* to 10, set *every* to 7 and *unit* to day, you will display the movement of solar objects the next ten weeks from the set date and time.

With the radio buttons you can choose which solar system objects to display, if, and how you want to display labels for these objects.

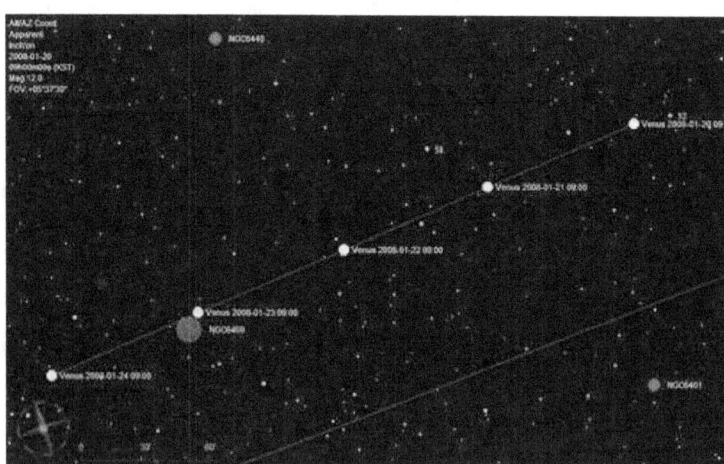

Animation

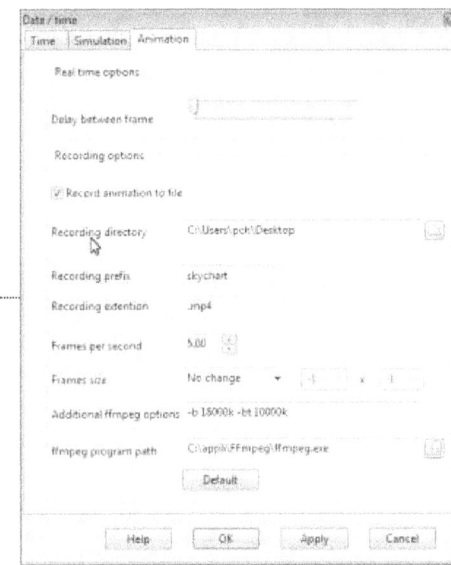

Real time options

Adjust the "Delay between frames" cursor to slow down the screen refresh during the animation. Put the cursor to the right for a two seconds delay or to the left for no delay. It is recommended to put it to the left when recording a movie.

Recording options

You first need to install the ffmpeg program to create the movie. For more detail see the installation section for your system. Set "ffmpeg program path" to point where you install this program, or just the program name if it is within your search path (default on Linux).

Check "Record animation to file" to make a video movie. Otherwise no file are saved and the animation is just played in the current chart.

Change "Recording directory" to indicate where you want the resulting movie. The default value is not convenient as this is under an hidden folder.

"Recording prefix" is used to name the movies. Using the default, the first one will be skychart1.mp4, then skychart2.mp4, and so on.

Set "Recording extension" to change the default ffmpeg container. See ffmpeg documentation [http://www.ffmpeg.org/documentation.html] for more details.

Adjust the "Frames per second" you want for the final movie. Use a low value from 0.5 to 2 if you want the same effect as the real time animation. Use a value between 15 and 30 if you want a smooth movie, but use a smaller time increment in this case. This set the -r option of ffmpeg.

Set "Frames size" to the size you want for the movie. The default is to not change and use the size of the current chart. If you set another size the chart will be resized when the animation start. This not set any ffmpeg option.

"Additional ffmpeg options" let you any option you want to give to ffmpeg. The default value set a relatively high bit rate to give a good quality result for all the preset size and fps. You can use this field to change the bit rate, but also the video codec. See ffmpeg documentation [http://www.ffmpeg.org/documentation.html] for more information.

You can try this options from a command line window before to put them here. The image sequence from the last animation run is keep until a new animation is run. So you can use them to try new option or even use another software to assemble them. You will find the files in the tmp folder under the User settings directory.
The default command is "ffmpeg -r 10 -i %06d.jpg -b:v 18000k -bt 10000k skychart1.mp4"

It is recommended to use the VLC [http://www.videolan.org/vlc/] software to visualize the resulting video without boring with the codec installation.

Date/Time settings are related to:

Chart Coordinate System. When you want to see what is visible above your horizon at a certain time and want to see time simulation in effect, the Chart Coordination System needs to be set to Azimutal Coordinate. There are several ways to achieve this. For example, from the menu bar: **Chart → Chart Coordinate System → Alt-Az Coordinate**

Solar System Probably you want to simulate movement of planets, asteroids or comets in whatever coordinate system. You need to set them visible, also here there are several ways. From the menu bar: **Chart → Show Objects** and check "Show

Planets", "Show Asteroids" and "Show Comets". Or the fastest way, by clicking the appropriate icons from the object bar in **object group B**.

Observatory Settings

From the menu: **Setup → Observatory Settings**

This dialog box can be reached from the menu by: **Setup → Observatory**, by the 🔬 icon in the **bar at the left hand side**, or by **Setup → All configuration options → All configuration options**.

The Observatory settings dialog box has two tabs:

- Observatory
- Horizon

For a proper display of objects on the chart, don't forget to check your **date / time** settings as wel.

Observatory

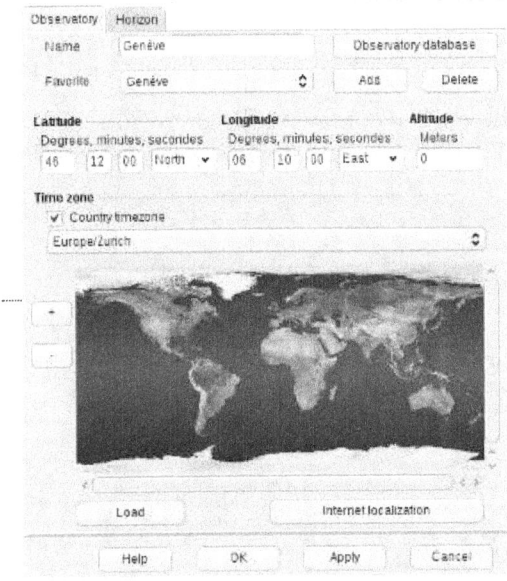

Define the location of the observation site to be used when calculating the altitude and the azimuth position of objects in the chart, rise and set time, solar system objects parallax, ...

There are several ways to enter your location, but it is advisable to start with the "Internet Location" button which, if your computer is connected to the network, will found at least your country. The relevance of the result varies greatly across countries and network providers.

You can then refine the result with the "Observatory database" button which allows you to choose among several millions of place, see below for details.

You can also enter the name of your location and click the map to get an approximate position. Or if you want the optimal precision enter the coordinates and altitude indicated by GPS receiver.

It is important to select the correct time zone for your observatory because SkyCharts needs it to calculate the *UT* from your *Daylight Savings* setting. This is very important to do proper ephemeris calculations in order to display the right chart.
It is recommended to use the country time zone as it correct for DST for any epoch where the rules are know.
You can also use an UTC time zone if you want a fixed time along the year.

If you frequently use this observation site you can add it to your favorites list with the Add button after all settings are right for you.

Observatory database

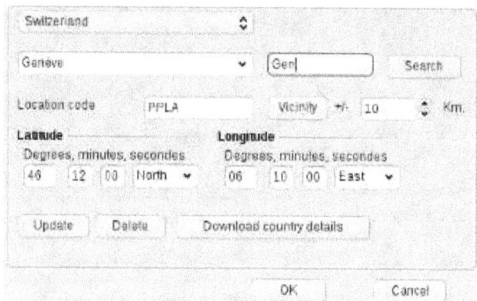

- Select your country
- Select your site from where you observe
  - To improve your choice, download the detail file of your country. If you want to copy this data to another computer without Internet access you can find them in the user data directory under the "tmp" folder.
  - Enter (a part of) the name of a near city in the search input box
  - Click the "Search" button
  - Select this near city from the pick-list in
  - Click the "vicinity" button
  - Now you can find names of places close to the city you previously selected in. Choose the one that suits you best.
  - The Location code help to solves homonyms place, such as a mountain and a river with the same name. Click in the box to view the list of codes.
- If the above is not sufficient, you can also add, modify or delete sites with "update" or "delete" button. To add a site, enter the name in the list and coordinates and click on "update". Use a prefix (i.e obs_) to quickly retrieve your favorite observing place by entering this prefix to the search box.

Horizon

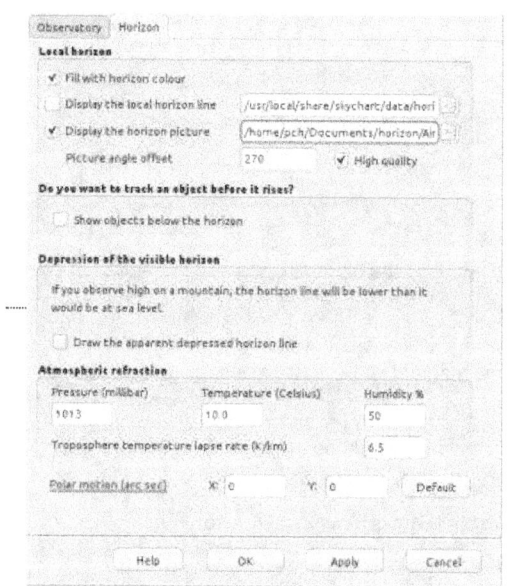

When your chart is set to use the Alt-Az coordinate system, you can display your local horizon as a line or as an area. To read more about changing the coordinate system, click **here**.
You can load a local horizon file by setting the path.

Write your own horizon file

You can write a file with a simple ASCII-editor like Wordpad or Vi to define your local horizon. As an example, you can open the file *[installation directory]/data/horizon/horizon_Geneve.txt*. As you can see from the file, the horizon is defined by a serie of records. Every line contains a pair of two values. The first value is the 'azimuth', the second is the 'altitude'. The units are degrees, where azimuth 0° is North and 90° is the Zenith. A dot (.) can be used as a decimal separator. You can put comment in your horizon file by lines that start with a mesh (#) character.

Use an horizon picture

You can also use a picture for the horizon panorama. The image must be a PNG or BMP file of any size representing the full 360° panorama with an equirectangular projection.
The sky area must be set transparent for files in PNG format or set the color to magenta (#FF00FF) for the BMP format.
The horizon must be exactly at the middle height of the picture, but you can cut the unused part. For example if the highest

point of your horizon is at an altitude of 20°, it is sufficient to have the picture to cover from -20° to +20°. Above +20° the sky will be fully transparent and below -20° the map is filled with the configured horizon color.

The left side of the picture is the East direction. If you use another orientation you must also give the offset angle. This angle can be read from the value of "angle_rotatez=" in a landscape.ini file found in the same folder as the picture.

If "High quality" is checked a point is computed for every screen pixel of the map. Otherwise it compute a point only every four pixel. See the discussion about the performance below.

This function is compatible with the horizon made for Stellarium [http://www.stellarium.org] using the Single Panorama Method [http://www.stellarium.org/wiki/index.php/Customising_Landscapes#Single_Panorama_Method].

So the quickest way to test it is to get a landscape file [http://www.stellarium.org/wiki/index.php/Landscapes] for Stellarium. Just be sure the format is compatible by looking at the row "type = spherical" in the landscape.ini file.

For example download the Jungfraujoch panorama [http://www.waldvogel.com/stellarium/landscape_jungfraujoch.zip], unzip the file in your home folder.

Launch Skychart and open the observatory setting, add an observatory location for Jungfraujoch as show here (location is not automatically set from landscape.ini but you can look at the file to get the right value):

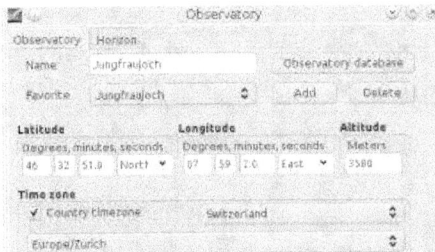

In the horizon tab, check "Display the horizon picture", click the folder button and open the file jungfraujoch-4096.png.

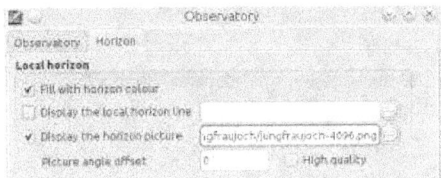

Return to the Observatory tab and click the "Add" button to add to the favorite list with all the picture options.

Click OK, you must see the following map:

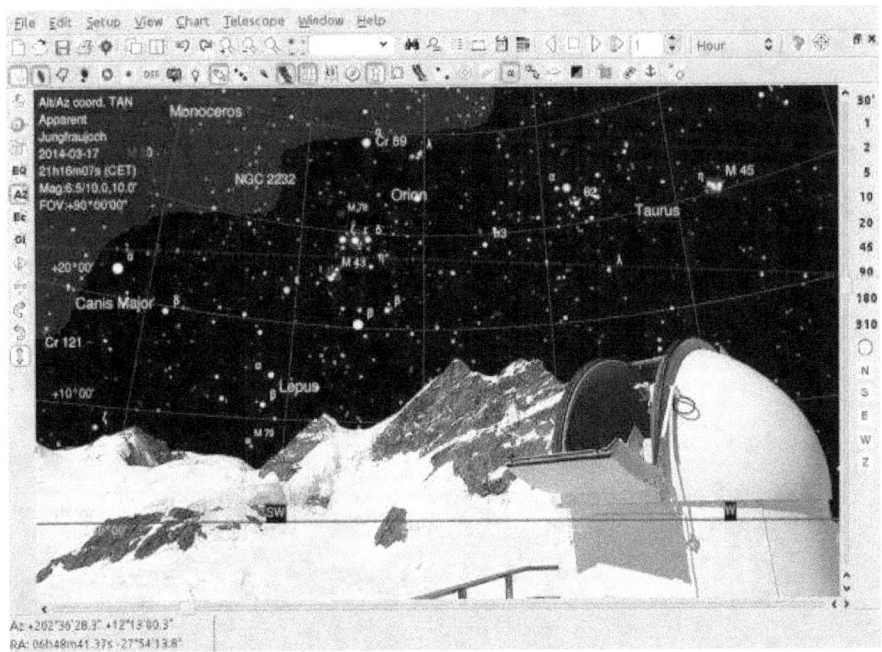

About the performance

On some situation the chart refresh performance can be very bad. The reason is because Skychart use only the main processor to display the chart, so if you have a big screen and a slow processor you can run into a problem.

Using a mid-range computer with a Core I5-2500 processor and a 1600×1200 screen with the chart set to full-screen, it take 0.3 second draw the Jungfraujoch panorama.

If you not get this value you can try the following:

- Uncheck "High quality" so the processor as four time less work to do. The quality must be enough if you use a high dpi screen (laptop with high resolution).

- Reduce the size of the chart window. With the same processor it take only 0.1 second for a 1024×768 window.
- Be sure you check "Reduce details when moving the chart" in Setup / Display.
- Try to cut the unused part of the picture. The Jungfraujoch panorama can be cut to 4096×1048 without loss.
- If you use a large picture (more than 4096 pixels) it can take a long time to load when you start Skychart. In this case try to use the BMP format that use less resources.

To make your own panorama

Put a camera on a tripod at the exact location you will put your telescope mount later. If you use a fixed pier, fix a photographic head on the pier. Be sure to level the tripod carefully. Try to have the camera objective near to the head rotation axis to avoid parallax problem on nearby objects.

Select a moment you get the most uniform lighting in every direction. A uniform high cloud cover can be good.

Get a sequence of picture for the whole horizon, be sure to have enough overlap between the pictures.

Use a panorama software like Hugin [http://hugin.sourceforge.net/] to assemble the pictures. Convert the result to PNG format to support transparency. Then use GIMP [http://www.gimp.org/] to set the sky transparent using one of the numerous method [https://www.google.com/#q=gimp+mask+transparency] available, I found that creating a sky mask first work fine.

Finally reduce the panorama size to a reasonable value, 4096 or 8192 pixel width.

If you observe from a mountainous area I also get success by using an online panorama generator [http://www.udeuschle.de/Panoramen.html]. In the form set your coordinates and "Left edge=90", "Right edge=90", "Tilt=0", "Elev. exaggeration=1". Show the panorama, save all the tiles and assemble in GIMP or using the "montage" command from ImageMagick.

The other possibilities

The other possibilities will be obvious from their description.
* Maybe you want to display objects below the horizon line.
* If your site is high on the mountain, maybe you want to simulate the horizon as a depressed line.

Above all, you can specify temperature, air pressure and humidity. This allows SkyChart to calculate and compensate for atmospheric refraction.

The last line is about some small correction for the Earth pole orientation. You can find the required prediction data in the latest IERS Bulletins A [http://www.iers.org/IERS/EN/Publications/Bulletins/bulletins.html] :

```
             MJD      x(arcsec)   y(arcsec)   UT1-UTC(sec)
2014  1 31  56688      0.0245      0.3483      -0.12827
```

For example for 2014 January 31 use X=0.0245 and Y=0.3483

Chart, Coordinates Setting

From the menu: **Setup → Chart, Coordinates**

The Chart, Coordinates Setting has six tabs:

Chart, Coordinates

Chart Setting

Here, you can specify the coordinate system used for the chart. The four available systems are:

- Equatorial coordinates
- Azimuthal coordinates
- Galactic coordinates
- Ecliptic coordinates

Type of coordinates

Precession, nutation and aberation recognition Your choice here determines which type of coordinates SkyCharts will display for a selected object on the status bar, on the bottom of the chart. If you don't check the **Expert mode**, you can choose in the groupbox 'Type of coordinates from four options·

- **Apparent, true equator, equinox and epoch of the date**
  This is the true apparent position for the equinox of date of the chart, including correction for proper motion, precession, nutation, aberration, light deflection. Normally you want this setting as this is the only to show the real position of the objects.
- **Mean of the date, mean equator, equinox and epoch of the date**
  The position for the equinox of date of the chart, including correction only for proper motion and precession. Use this setting only to compare the coordinates with other source that use this system, like printed almanac
- **Mean J2000, mean equinox and epoch J2000**
  The mean position for the date 2000.0, including correction for proper motion and precession at 2000.0. Use this setting only to compare the coordinates with other source that use this system, like printed atlas.
- **Astrometric J2000 , mean equinox J2000 and epoch of the date**
  Use the equinox 2000.0 and precession for 2000.0, but proper motion for the current chart date. This setting is only used to compare the coordinates with an astrometry software.

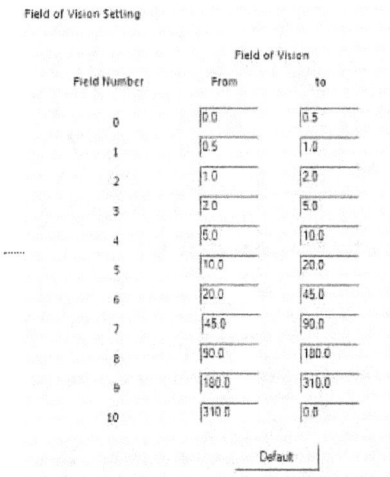

Chart Setting

Chart Coordinate System

- Equatorial Coordinate
- Azimuthal Coordinate
- Galactic Coordinate
- Ecliptic Coordinate

Equinox: 2000.0   Epoch: 0.0

Stars proper motion options (if available in the catalog)
☑ Use the proper motion to correct the position for the given epoch

Nutation · Aberration
- Mean position
- True position

☑ Expert mode

In **Expert mode** you can specify more details :

- **Equinox** year (from -20000 to 20000)
- **Epoch** year (from -20000 to 20000)
- **Proper motion** checkbox to correct positions for the given epoch
- **Mean position** or **True position** for *Nutation* and *aberration*.

Fast ways to make simple changes to the chart coordination system are from the menu by **Chart → Coordinate Sytem → [Your Choice]**, or directly on the chart by the icons in **coordinate system group**.

Field of Vision

Field of Vision Setting

Field Number	Field of Vision	
	From	to
0	0.0	0.5
1	0.5	1.0
2	1.0	2.0
3	2.0	5.0
4	5.0	10.0
5	10.0	20.0
6	20.0	45.0
7	45.0	90.0
8	90.0	180.0
9	180.0	310.0
10	310.0	0.0

Default

Here you can specify eleven ranges as the Field of Vision (numbered from 0 to 10).

For each range you can specify the minimum FOV in degrees, which automatically will become the maximum for the preceding range.

The first range minimum is 0.0° and last range maximum is 360° (which funny enough appears as 0.0°), these values cannot be modified.

These ranges are listed at bottom of all the **Catalog** dialog box tabs and also are used with the **Projection**, **Object Filter** and **Grid spacing** tabs.

Changing the FOV itself can be done from the menu by **Chart → Field of Vision** or directly on the chart by the icons in **the field of vison group**. A very precise FOV can be manually set by the FOV part in the dialog box from the menu **View → Field of Vision (FOV)**.

Projection

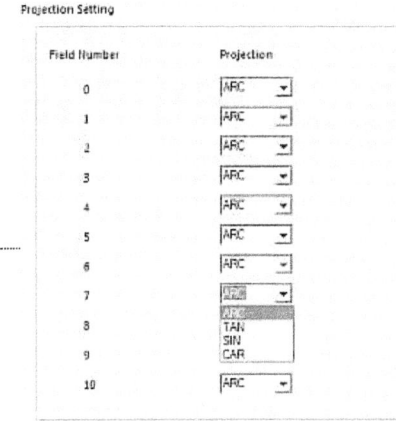

For every FOV range you can choose from four projection types:

- **ARC** Zenithal equidistant. It corresponds to the projection of a Schmidt camera.
- **TAN** Gnomonic. This is the default projection up to 90°. Corresponds to the projection of a picture obtained with a telescope or a photographic lens. The tangent projection has a great distortion for fields larger than 90° and diverge for 180°
- **SIN** Slant orthographic. Used to display images in radio-astronomy. The sine projection overlaps at more than 90°.
- **CAR** Cartesian. It is of no great interest, apart from the fact that it can display very large field of views.
- **MER** Mercator. A good projection for large fields but diverge at the pole. This is the default projection between 90° and 360°.
- **HAI** Hammer-Aitoff. A compromise often use for large fields.

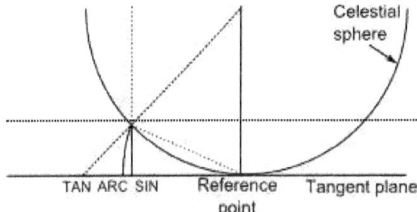

The three zenithal projections by E. Griessen, AIPS memo 27

By default the projection CAR, MER and HAI are oblique, i.e the projection equator is moved at the chart center. This minimize the distortion on the chart.
If you want the classic aspect of this projections you need to remove the corresponding check box.
See the projection comparison page for more information.

Object Filter

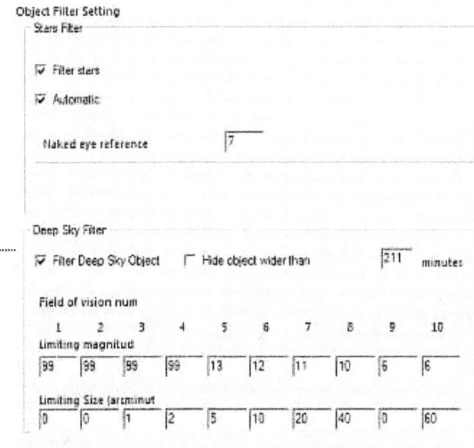

By this tab you can configure magnitude limits to display stars and deep sky objects based upon the FOV of your chart.

The **Stars Filter** can be:

- disabled. (only practical with the smaller FOVs),
- automatic. By this you can specify a magnitude as **Naked eye reference**,
- manual. Now you can specify a magnitude limit for every FOV range.

A funny excercise: There's no technical problem to disable the stars filter in combination with a large **active star catalog** (i.e. HST GSC), a high value for *field number max* (i.e. 6) and a **FOV** of 20 degrees. Now you can see why a disabled Stars filter in a large FOV is not very convenient.

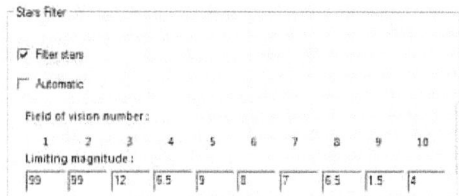

Deep Sky Filter can be:

- disabled (only practical with the smaller FOVs),
- manual. Here you can specify the magnitude limit for every FOV range.

You can also filter large deep sky objects specifying the maximum dimension in minutes.

Grid Spacing

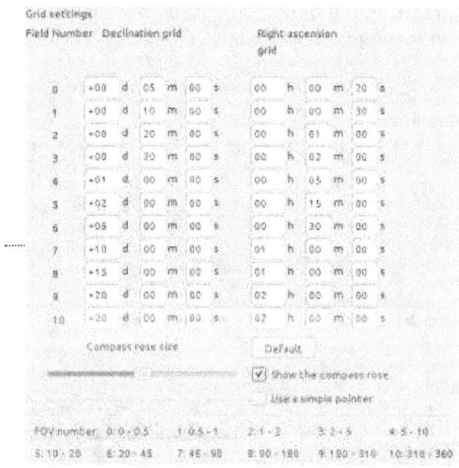

By this tab you can set the grid spacing for every FOV range.

- **Degree** is for the spacing altitude or declination direction
- **Hour** is for the azimuth or right ascension direction

You can enable or disable the grid display for every FOV range.

You can enable or disable the display of a compass, you also can adjust its size or disable it.
You can use a simple N/E pointer instead of the compass.

You can enable or disable the display of grids from the menu by **Chart → Lines / Grid → [Show coordinate grid/add equatorial grid]**

Object List Setting

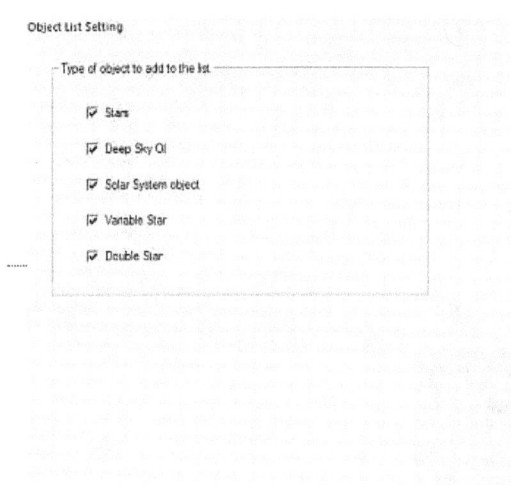

Object List Setting

Type of object to add to the list
- ☑ Stars
- ☑ Deep Sky Ol
- ☑ Solar System object
- ☑ Variable Star
- ☑ Double Star

By this tab, you determine which object types are to be filtered from your chart to your **Object List**. Click on the ☷ icon from the **main bar** to retrieve your filtered list of objects displayed on the chart.

Projection comparison

Large field of view

The larger field of vision make the more visual difference between the different projection.
The example show a field of 150° and 360° wide using the different options available in the menu Setup -> Chart, Coordinates

MER, Mercator

150° FOV, standard and oblique :

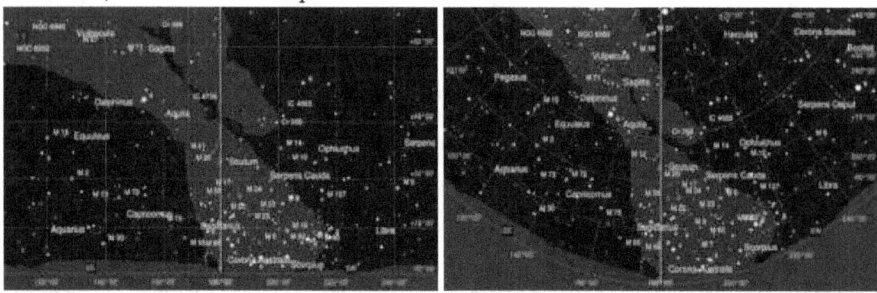

360° FOV, standard and oblique :

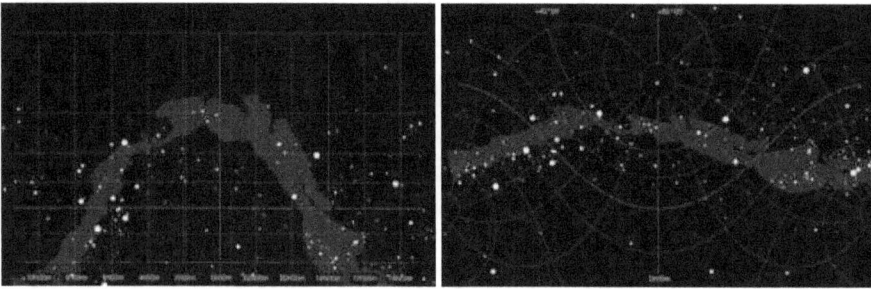

HAI, Hammer-Aitoff

150° FOV, standard and oblique :

360° FOV, standard and oblique :

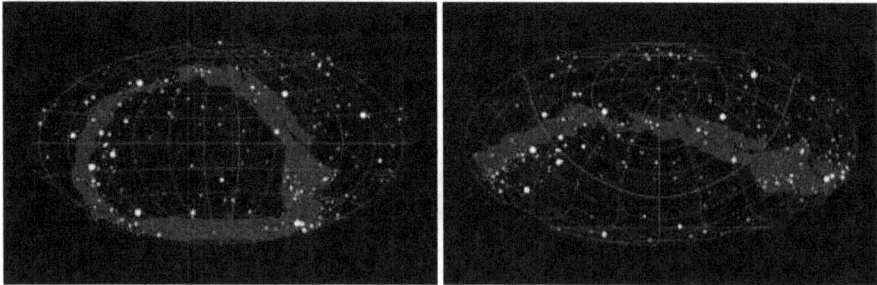

CAR, Cartesian

150° FOV, standard and oblique :

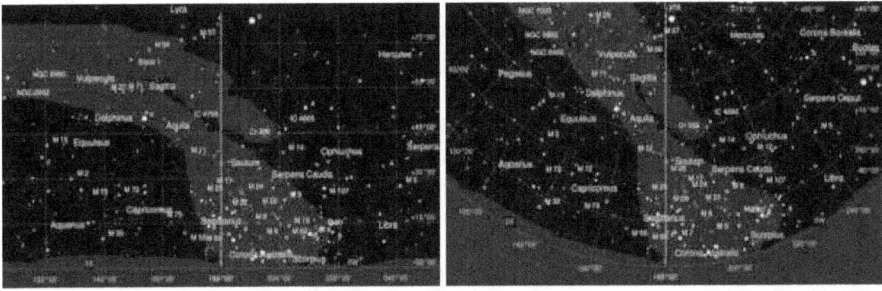

360° FOV, standard and oblique :

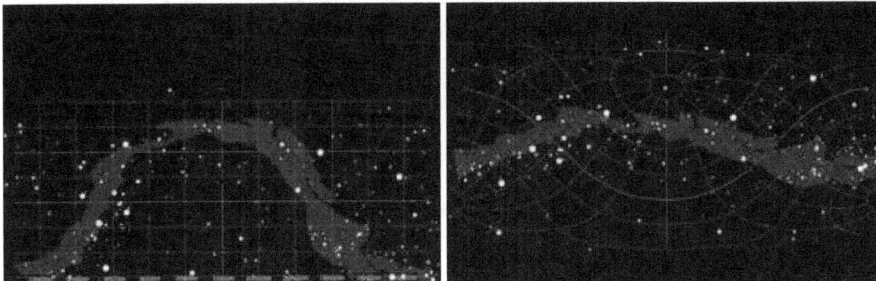

ARC, full sky

When you click the "Show all sky" button⊙ the projection automatically change to the ARC full sky show below.
This same change of projection occur if you use simultaneously: an oblique projection, an azimuthal coordinate system, a very wide field of vision, and the altitude at the chart center is greater than 45°.
It return to your selected projection if you zoom in or move the chart center nearer the horizon.

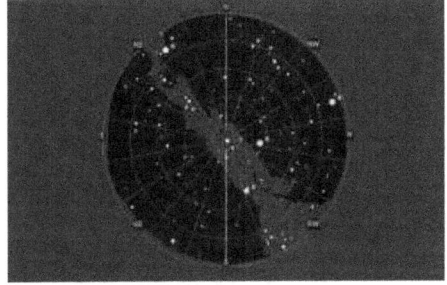

Small field of view

On smaller field of vision the difference between the projections is more subtle but important if you want to match a specific image.
The example here show a FOV of 70° to make some difference visible at this scale, look at the grid curvature for more evidence.

TAN, Gnomonic

70° FOV :

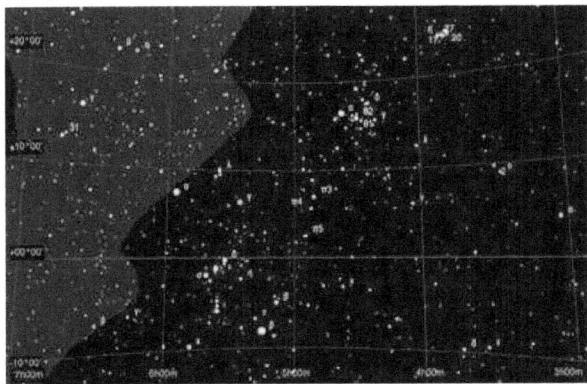

ARC, Zenithal equidistant

70° FOV :

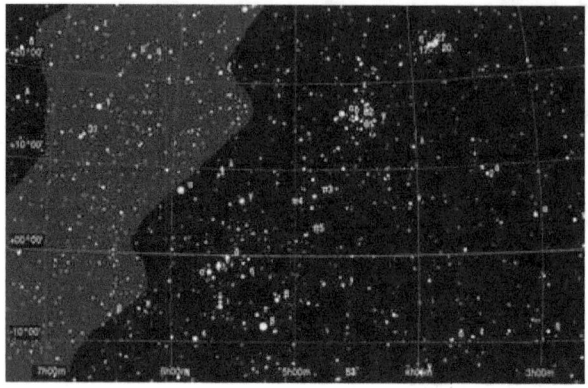

SIN, Slant orthographic

70° FOV :

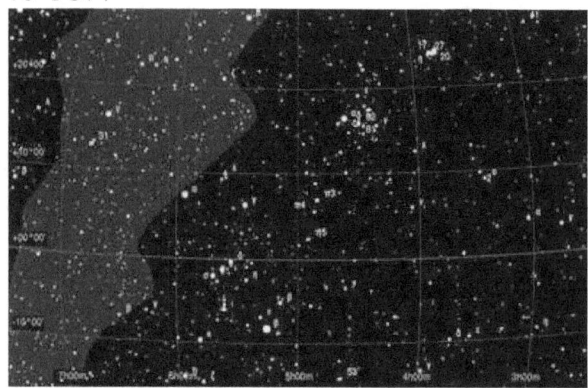

MER, Mercator

70° FOV, oblique :

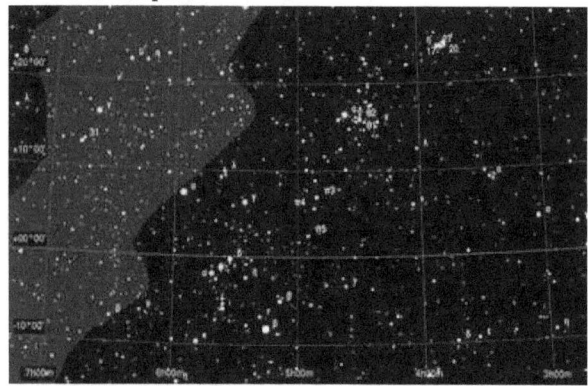

HAI, Hammer-Aitoff

70° FOV, oblique :

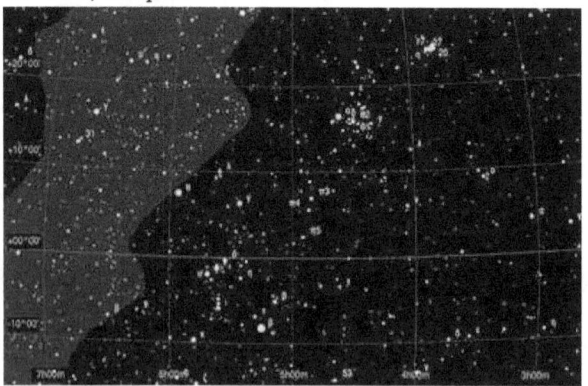

CAR, Cartesian

70° FOV, oblique :

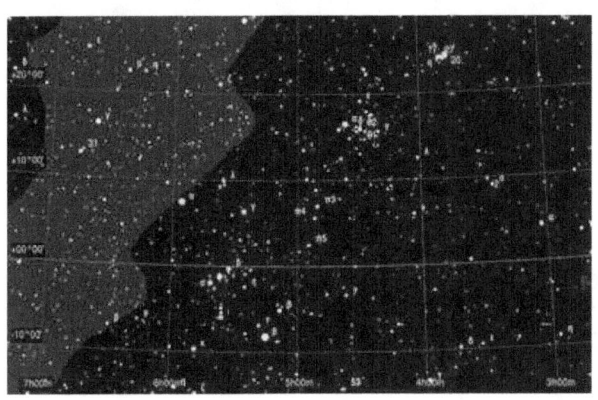

91

Catalog

From the menu: **Setup → Catalog**

Catalog

From the menu: **Setup → Catalog → Catalog**

Here you can add and activate all kinds of external catalogs previously built for SkyCharts with the CatGen utility. You can invoke Catgen by the *CatGen* button at the top right. Read more about the new creation or adaptation of existing catalogs for SkyCharts with CatGen here.

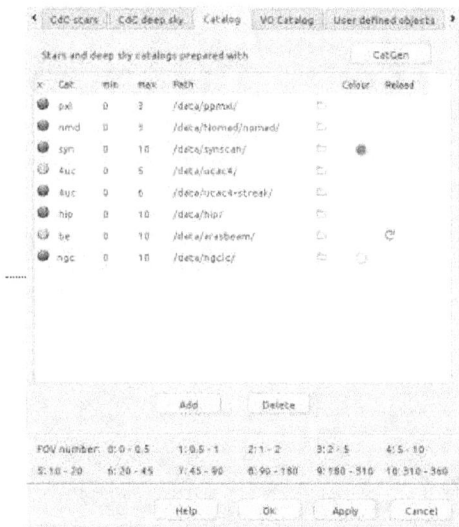

Before you can add a new catalog, you need to add a line where you will describe its configuration. (You could also remove a selected line by pressing the "Delete" button.)

Next, you need to set the path to your new catalog. Click the directory icon at the right of the line to set the path. You can have your directory with your external catalog anywhere on your computer, but it is generally a good idea to place them together as subdirectories of the `cat` directory in the SkyCharts installation directory. In the directory of your catalog there must be a `.hdr` file whith the description of this catalog. Select the `.hdr` file. In case of an error, the first cell on the line is boxed in red.

In the **min** field, enter the **field of vision number** to set the smallest fields of vision that you want your catalog to be used for in SkyCharts. (For your current FOV number settings, have a look at the bottom of the dialog box.) A good choice to start with can be to put a **0** here, otherwise zooming in can result in the loss of objects that were visible in a larger Field of vision. In the **max** field, enter the largest field of view number for which you want to use this catalog.

If the catalog allow for this option you can change the color of the object by a click on the icon in this column. Select a black color to return to the default color.

Some catalog allow for an update URL to reload the catalog data from the Internet. Click the circular arrow in the Reload column to download the latest version of the catalog.

Click the red cell at the left of the line to activate the catalog. If all is set properly, it will switch to green.

VO Catalog

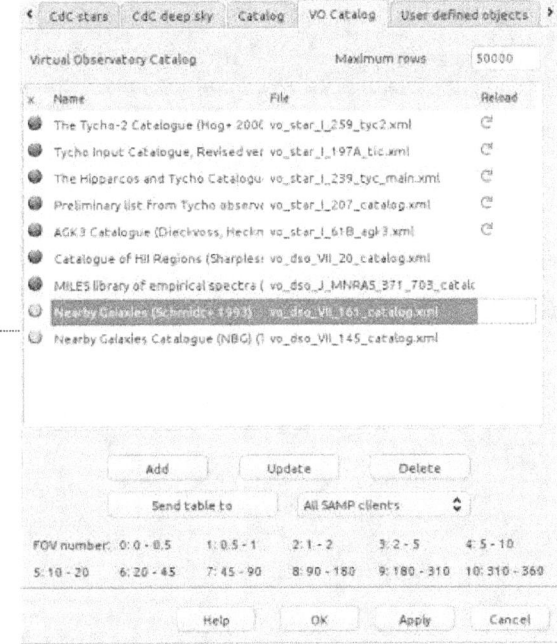

From the menu: **Setup → Catalog → VO Catalog**

By this tab you can manage any online catalog available with the Vizier Virtual Observatory interface, there are more than 9000 catalogs

This requires an active connection to the Internet to search and download a new catalog. But after you acquire the data you also can use them offline. You can copy the files to another computer, they are located in the User settings/vo directory. Locate the file name in the **File** column and copy both the .xml and the .config files.

Set **Max rows** to the maximum number of records you want to download at a time for a catalog. You can increase the default value if you want a full catalog to use offline. Or decrease the value if you suffer of slowdown when the chart is redraw. Any change takes effect with the next download.

Initially the list is empty, click the **Add** button to launch the Virtual Observatory interface and follow the instructions on this page.

If you want to change the settings for a catalog, like the column selection, the drawing symbol or the color, select the catalog in the list and click the **Update** button.

The **Delete** button removes all the associated files for the selected catalog.

If you are connected to a SAMP hub you have the option to send the selected table to one or all of the other clients.

Click the first cell at the left of the line to make it green and activate the catalog. Switch the cell to red when you want to disable this catalog.

To plot all this data to the map, the button on the top bar needs to be checked. This is a quick way whether to select or not all these Virtual Observatory catalogs.

If the catalog is not downloaded completely, as for a large survey that includes millions of objects, you can click the arrow at the **Reload** column to actualize the data for the current chart field.

Select first a narrow enough field of view on the chart, 1° or 30' is a good start. Then right click the button to open this window and reload the catalog you want.

User defined objects

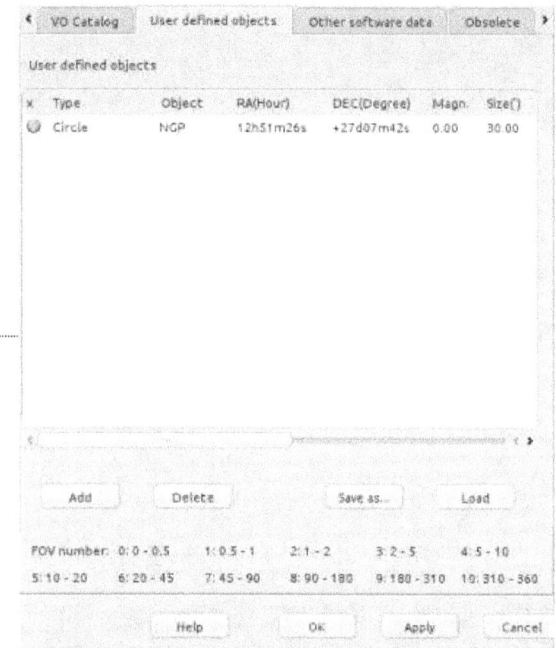

From the menu: **Setup → Catalog → User defined objects**
You can add here any object or point of interest you want to display on the map.

Click **Add** to add a new object, and select it's properties. At least the coordinates must be entered.

Then click the red dot on the left to make it green to select the object for display.

To plot all this data to the map, the button ＊ in the top bar need to be checked. This is a quick way to select or not all the objects.

You can also save and load the list of object to a file using the corresponding buttons.

CdC Stars

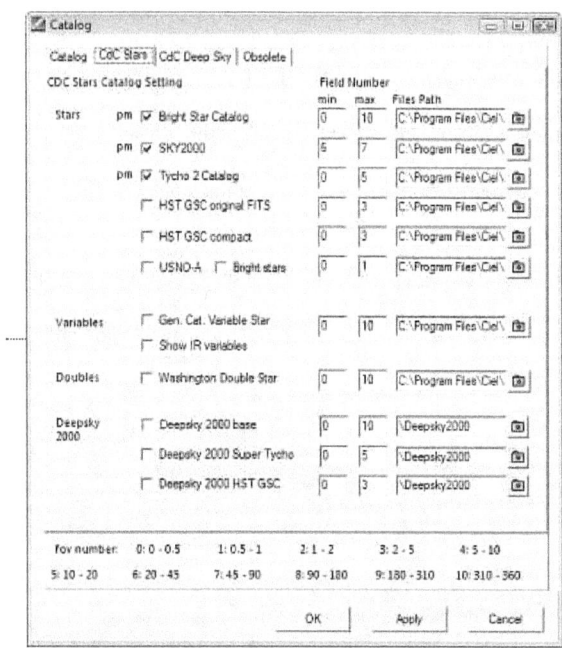

From the menu: **Setup → Catalog → CdC Stars**
By this tab you can activate SkyCharts to display the stars from your catalogs. Simply place a check in the appropriate checkboxes. In the standard situation, SkyCharts will search the catalogs in their preconfigured paths where SkyCharts assumes the appropriate catalogs reside. If SkyCharts can't find the catalog in the configured file path, the file path input area will display a red colour. You need to set this right before SkyCharts can use your catalog.
Don´t panic if you can´t find the proper path. Because of their sizes, not all catalogs came with the SkyCharts installation, maybe you need to install the catalog. Click here for further reading about installation of extra catalogs. The Deepsky 2000 is

a package of catalogs that isn't available for free, you can purchase it from **here [http://www.deepsky2000.net/]**.

In the **min** field, enter the **field of vision number** to set the smallest fields of vision that you want your catalog to be used in SkyCharts. (For your current FOV number settings, have a look at the bottom of the dialog box.) A good choice to start with can be to put a **0** here, otherwise zooming in can result in the loss of objects that were visible in a larger Field of vision. In the **max** field, enter the largest field of view number for which you want to use this catalog.
If you are using multiple catalogs, you can enhance the performance of SkyCharts by avoiding the usage of multiple catalogs in the same field of view.

Catalogs marked with "pm" contain "proper motion" information. The program uses the proper motion to calculate the position of the object for the configured **date**. It is possible to display the proper motion by showing the displacement for a period of 1-9999 years in the form of a line. See **Display Lines**.

You can configure how the stars are displayed by the tab **Color** from the menu by **Setup → Display**. You can switch the display of stars on or off by pressing the ☼ icon of the **object bar**.

CdC Deep Sky

From the menu: **Setup → Catalog → CdC Deep Sky**
By this tab you can activate SkyCharts to display the deep sky objects from your catalogs. Simply place a check in the appropriate checkboxes. In the standard situation, SkyCharts will search the catalogs in their preconfigured paths where SkyCharts asumes the appropriate catalogs reside. If SkyCharts can't find the catalog in the configured file path, the file path input area will display a red colour. You need to set this right before SkyCharts can use your catalog.
Don´t panic if you can´t find the proper path. Because of their sizes, not all catalogs came with the SkyCharts installation, maybe you need to install the catalog. Click **here** for further reading about installation of extra catalogs.

In the **min** field, enter the **field of vision number** to set the smallest fields of vision that you want your catalog to be used in SkyCharts. (For your current FOV number settings, have a look at the bottom of the dialog box.) A good choice to start with can be to put a **0** here, otherwise zooming in can result in the loss of objects that were visible in a larger Field of vision. In the **max** field, enter the largest field of view number for which you want to use this catalog.

For a better performance and to avoid duplicates of objects to be displayed, you better do not use multiple catalogs for the same type of deep sky objects in the same field of vision. As indicated at the bottom of the tab, it is recommended to use catalogs from only one of the three catalog groups. You can choose to mix those groups, but because of the duplicates as a result of the overlapping a warning message will be displayed.

You can choose how to display the symbols of deep sky objects by the tab **Deep-sky Colour** by the menu **Setup → Display**. You can switch the display of deep sky objects on or off by pressening the ☼ icon of the **object bar**.

Obsolete

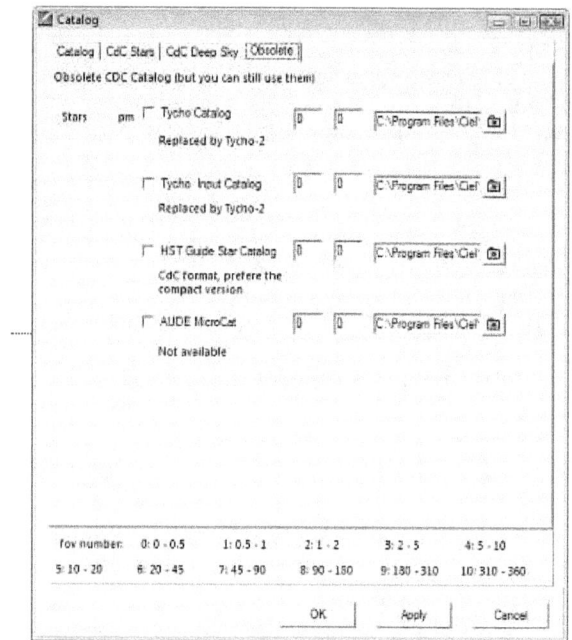

From the menu: **Setup → Catalog → Obsolete**

If you want to use your obsolete catalogs, specify them in this tab the same way you did in the **CdC Stars** and **CdC Deep Sky tabs**.

Virtual Observatory interface

From the menu: **Setup → Catalog → VO Catalog → Add**

This screen enables you to select and download data from more than 9000 catalogs available from the Vizier [http://vizier.u-strasbg.fr] Virtual Observatory interface.

Catalog selection

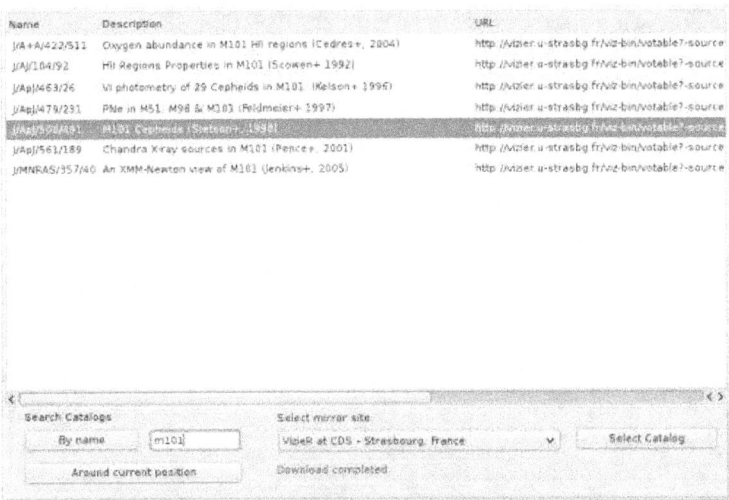

The first time you open this window you need to select a **Vizier mirror** near your location.

There are two ways to get a list of the catalogs:

- **By name** : type some words of the catalog name, an acronym, an object names, or the survey name.
- **Around current position** : this will search for catalogs with data around the current map center coordinates.

Scroll the list to select the line with the data you want and press **Select catalog** to go to the next screen.
For an example, here we select the Cepheids in M101 galaxy.

Table and parameters

Be careful and take your time to examine this screen. There are some options that greatly affect the final result. Despite the program tries to determine the best settings from the data it receives, you probably want to alter a few for your display.

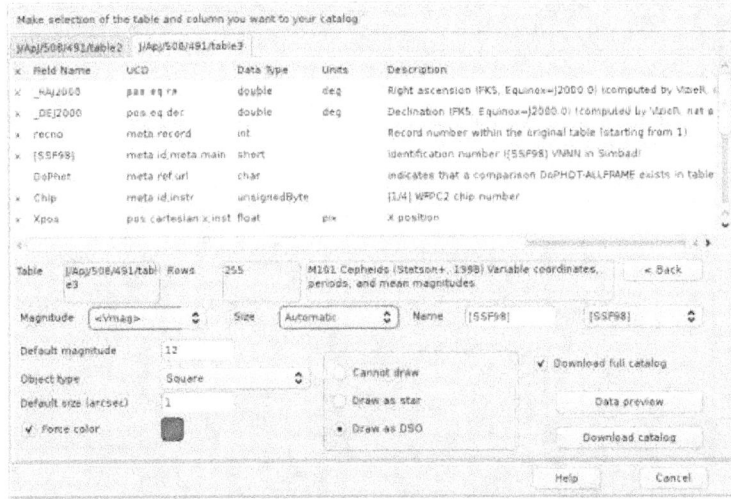

The next step is to click on the top tabs to select the data table you want from the catalog. Only the tables that contain the object coordinates are included because these are required to plot the objects on the map.
For example, here the table 2 lists the reference stars and table 3 the M101 Cepheids, so we select table 3.

The list contains all the fields of the table. With a "x" in the first column the field is selected to be included in our download. The program will use the UCD [http://www.ivoa.net/Documents/latest/UCDlist.html] field descriptions and the units to recognize the data. Review the list carefully to be sure the required data to plot the object is selected:

- identification (ucd = meta.id,meta.main)

- magnitude (ucd = phot.mag,em.opt.V)
- size (ucd = phys.angSize)

If the table contains more data of a similar type (same UCD), you can select below which column to use on the map for the magnitude, the size and the identification. You can also change the name prefix by another text than the column title.
The Units are also important, don't expect the program to plot the object at it's correct size if it is given in millimeters on the photographic plate!

At this stage it may helpful to have a look at the actual data. Click the **Data preview** button to load a sample in the screen described below.

Below the list are the table name, the total number of records, the table description and a check box to select if you want to download the full catalog or just the data at the current map position.

On the left you can select:

- The default magnitude to use if the magnitude data is missing.
- The object type or symbol to plot on the map
- The default size in arc seconds to use if this data is missing.

To override the default color for the type of object: Check "Force color" and click on the color square.

Then check the kind of drawing "as Star" or "as Deep sky object" in the middle box.
Sometimes the program makes a wrong choice. For example it wants to plot a star catalog as DSO because it includes some angular measurement wrongly taken as the object size.
Or as with the Cepheids in our example, we prefer to draw the star position with a symbol, so we select DSO.

When all is ready press the **Download catalog** button.

When it is completed, press the **Close** button to return to the catalog settings, or the **Back** button to return to the Catalog selection.

Data preview

This is just a preview of a few rows of the data you selected in the previous screen.

Press the **Back** button to return.

Result on the map

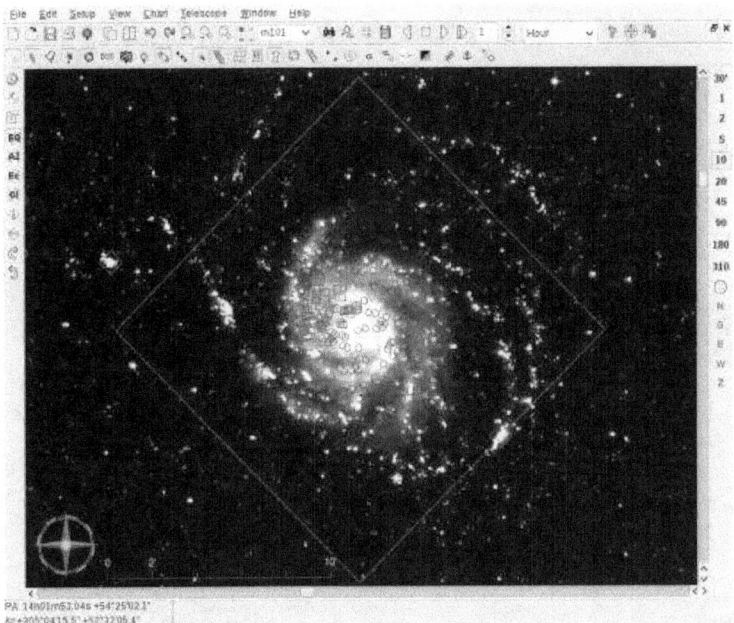

Our Cepheids example are the green squares.

We previously selected the M101 HII region as blue circles, the HyperLeda catalog as pink lozenges, and the stars from the NOMAD catalog.

Note that the VO button on the top bar is checked to plot this additional data.

If you need more powerful functions to select the data of interest you can use Topcat [http://www.star.bris.ac.uk/~mbt/topcat/] with the SAMP interface.

Solar System

From the menu: **Setup → Solar system**
You can make settings for the Solar system objects in four tabs:

- Solar System
- Planet
- Comet
- Asteroid

NEO's (Near Earth Objects) demand a special approach to have them displayed.

Solar System

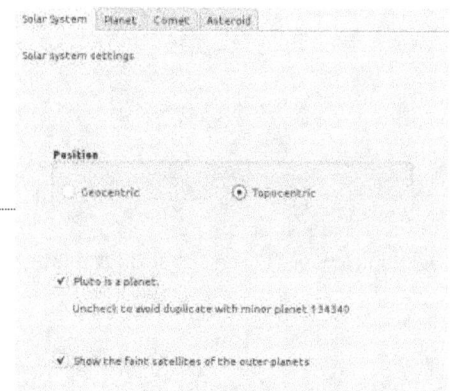

You specify to calculate geocentric or topocentric positions. Always keep it to topocentric unless you want to compare the data with a printed ephemeris.

Also you can choose if SkyChart should consider Pluto as a planet or not.

The last option is to show or not some 33 faint satellites of the outer planets. Most of this satellites are never visible with any telescope from Earth so it make sens to disable this option.

Ephemeris

To give you the best possible precision, the program uses the JPL DExxx ephemeris to compute the planet positions and the nutation.
To limit the download size, the included DE405 file with this software is limited for usage from the years 2000 to 2050.

You can add more files to cover an extended time range or to use other ephemeris. Download the binary files from
ftp://ssd.jpl.nasa.gov/pub/eph/planets/Linux [ftp://ssd.jpl.nasa.gov/pub/eph/planets/Linux] or
ftp://ssd.jpl.nasa.gov/pub/eph/planets/SunOS [ftp://ssd.jpl.nasa.gov/pub/eph/planets/SunOS].
Copy the linx* or unx* files to the installation folder under data\jpleph\ without creating a subdirectory. You must restart the program after adding or removal of files.

The software tries to load DE430, DE431, DE423, DE421, DE422, DE405, DE406, DE403, DE200 in this order.
The "lnx*" files are used in preference of the "unx*" because they do not require byte swapping.

If no file is found for the date, it uses plan404 by Steve Moshier [http://www.moshier.net/index.html] that allows for computation from -3000 to +3000 with a precision better than one arc second.

If DE431 is present the planet calculation can be done between -13000 and +17000.

If you select a date for which there is no mean to compute the planets position, the planet display is automatically disabled. You need to click the Show planet button again after you return to a valid date.

Planet

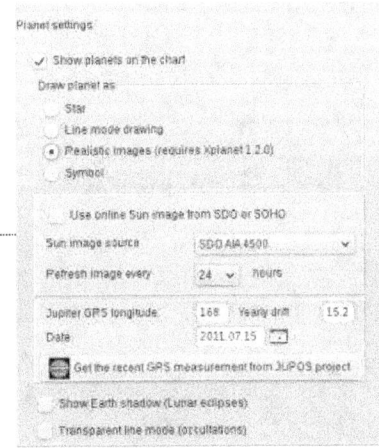

Here you can enable or disable the display of the Sun, planets and satellites on the chart.

Four modes to draw planets are available. You will notice the difference in fields of vision smaller than 2°.

Check the corresponding box to show a real time image of the Sun from the satellites SDO [http://sdo.gsfc.nasa.gov/] or SOHO [http://sohowww.nascom.nasa.gov/home.html].
Select the image you want from the list. The default AIA 4500 correspond to the visual aspect but you can also try the other wavelengths.
The image will be refresh after the selected time or each time you select another wavelength. Effective update frequency depend on the satellite operation, see the satellite web page for more information. The image source and time is indicated next to the Sun on the map.
This feature require an Internet connection and correct setting of the proxy server if required.

You can update the position of Jupiter's Great Red Spot (GRS) in longitude: Click at the icon at the right to open **Jupos page [http://jupos.privat.t-online.de/]** on your browser, where you can find the GRS longitude in the menu at the left. Look at the red dot at the bottom of the graph to get the most recent position of the GRS center. For example it was about 160 in December 2010:

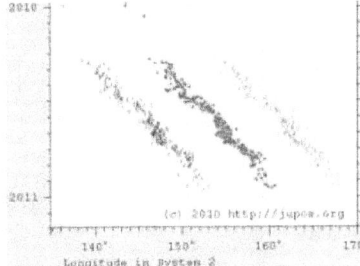

You can check the **Show Earth shadow** too. When you check this, the position of the Earth shadow will be displayed on the chart as two concentric circles, slightly darker than the background. The dimensions of the circles correspond to the size of the Earth shadow at the distance of the Moon, in order to simulate the lunar eclipses. The core shadow (umbra) is displayed in the middle, around it is the zone of partial eclipse (penumbra).

Transparent line mode allows the stars to be displayed behind planets or the Moon. This can be useful with occultations, to predict where the star will reappear.

Comet

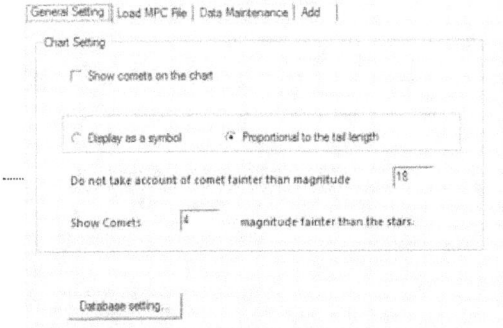

Before SkyCharts can display any comets, you must load a file with the orbital elements of the comets in the database on your computer. The usual way to achieve this, is simply to download the file of the latest version from the Minor Planet Center (MPC). You can also add new comet orbital elements, delete all entries, or selectively delete obsolete entries from the database. There are four tabs in this window:

- **General Setting** where you can choose to:
  - enable or disable the display of comets on the chart.
  - display the comet as a fixed symbol, or with a tail, sized proportional to its estimated magnitude.
  - filter comets below a limiting magnitude. You can display comets with a given magnitude fainter than the faintest displayed stars.

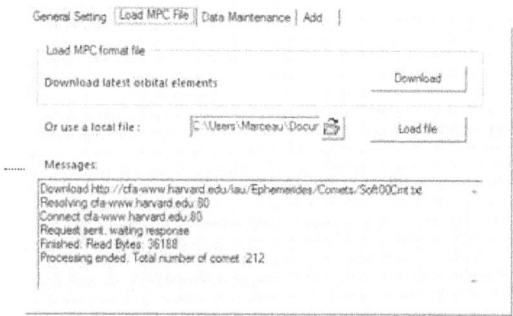

- **Load MPC File**
  - **Download** button to retrieve the last file with orbital elements for comets from **IAU Minor Planet Center [http://minorplanetcenter.net/]** and store the information in the database. The source file will be saved as COMET-yyyy-mm-dd.DAT, where yyyy-mm-dd is the date of download.
    If downloading is a problem, check the upper part of the **Setup → Internet → Orbital elements** dialog box.
  - **Or use a local file:** with its directory icon gives you the opportunity to load any COMETxxx.DAT in the database with **Load file** button. It will be used automatically after you load it.
  - A message window will display the processing steps and the results.

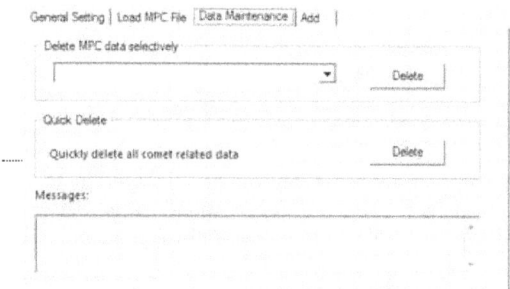

- **Data Maintenance** where you can delete obsolete entries in the database, or delete the all the comet information from the database. A message window displays the processing steps and the results.

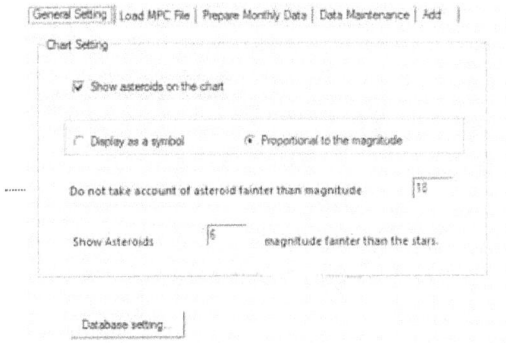

- **Add** where you can add orbital elements for a new comet in your database.

- Have a look here about how to enter the orbital element data.

Asteroid

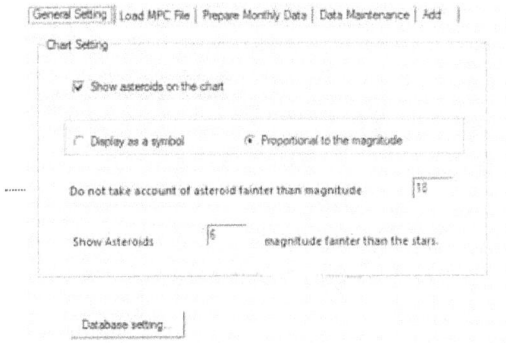

Before SkyChart can display asteroids, you need to have a file with the orbital elements of the asteroids. It is quite simple to download the data of the latest version from the Minor Planet Center (MPC) to use in the database on your computer. You can also add new asteroid orbital elements, delete all entries, or selectively remove obsolete entries from your database.

There are five tabs in this window:

- **General Setting** where you can choose to:
  - enable or disable the display of asteroids on the chart.
  - show asteroids as a symbols, or as stars, proportional to their magnitude.
  - filter asteroids below a limiting magnitude. You can display asteroids with a given magnitude fainter than the faintest displayed stars.

- **Load MPC File**
    - **Download** button to retrieve the last version file with the orbital elements of asteroids(bright, unusual and distant asteroids) from **IAU Minor Planet Center [http://minorplanetcenter.net/]** and store the information in the database. The source files will be saved in a file like: MPCORB-yyyy-mm-dd.DAT where yyyy-mm-dd is the date of download.
    If downloading is a problem, check bottom part of the **Setup → Internet → Orbital elements** dialog box.
    - with **Use a local file** and its directory icon, you can specify any MPCORBxxx.DAT and load it in the database with the **Load file** button. It is automatically used after download.
    - A message window will display the processing steps and the results.

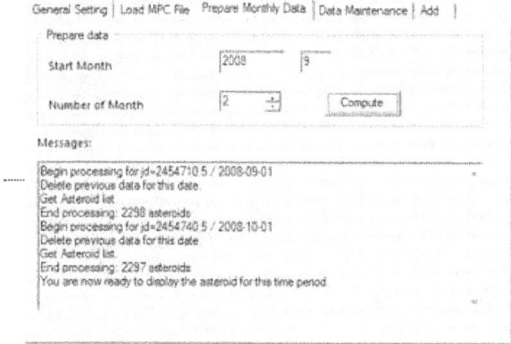

- **Prepare Monthly Data** process the orbital elements data to put it in the database for the given months range. A message window will display the processing steps and results.

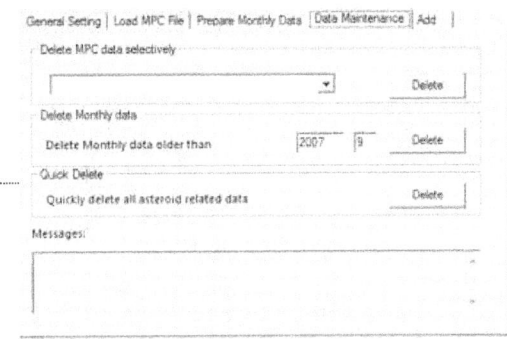

- **Data Maintenance** where you can delete a single obsolete entry. Or if you wish, all asteroid information older than a given month from the database, or all the asteroid information from database. A message window will display the processing steps and the results.

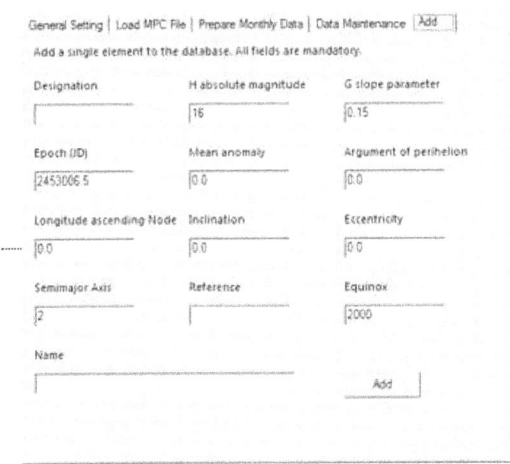

- **Add** where you can add orbital elements for a new asteroid in your database.

- Have a look here about how to enter the orbital element data.

Near Earth Objects

NEOs demand an entirely different approach to get their positions displayed with sufficient certainty. You will need to download calculated position data to build a catalogue using CatGen. For further details, please have a look here.

Display

From the menu: **Setup → Display**.

The Display setting has nine tabs:

- Display
- Color
- Deep-sky colour
- Sky colour
- Lines
- Labels
- Fonts
- Finder circle (Eyepiece)
- Finder rectangle (CCD)

Display

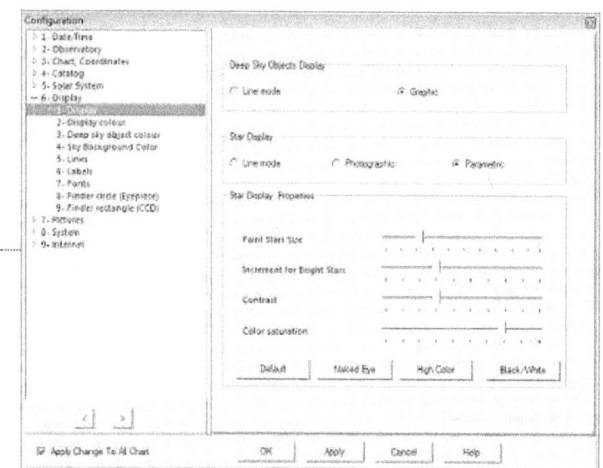

From the menu: **Setup → Display → Display**

The way stars and deep sky objects are displayed.

You can choose to display the deep sky objects by outlines (Line mode), or to display them as filled areas (Graphic).

The display of stars can be set to **Line mode** (a simple drawing, not very cosmetical), **Photographic** (a bit like a photograph, no controls) or to **Parametric**. In parametric mode, you have lots of controls like the size of stars, brightness, contrast and colour saturation. Four combinations of these parameters are built in: **Default**, **Naked eye**, **High color** and **Black/white**.

The "Anti alias" is normally checked on. You can try to deactivate it if you get slow performance on a old computer.

Color

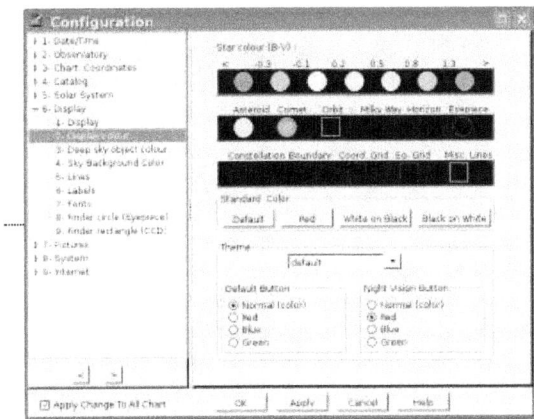

From the menu: **Setup → Display → Color**

In this menu, you can change the color of the objects, lines and grids on your chart. Simply click on the circle or square under its label. Then, choose the colour by a click in the coloured pop-up window, followed by a click on the "OK" button.
The top row is for stars, differentiated on their B-V magnitude ratio. The next rows show you solar system objects, and lines of all sorts. You can change the colours for deep sky objects in the next tab.
In the group-box **Standard Color** you can choose from predefined color-schemes: **Default** will display the colors on the chart like you would see them on real sky. **Red** will change to an eye-gentle colour), **White on Black** displays black objects on a white background, and **Black on white** displays white objects on a black background (the two last labels seem to be switched around, bug report 457).

By the groupboxes **Night Vision Button** and **Default Button** you can define the displayed color-themes when you activate or deactivate the **Night Vision Button** in the Main Bar.

You can configure which star catalogs are used by the tab **CdC Stars** from the menu by **Setup → Catalog**. You can switch the display of stars on or off by pressing the ⚫ icon of the **object bar**. The display of planets, comets, asteroids, ecliptic, equators, horizon and eyepiece can also be switched of or on by their appropriate icons from the object bar in the **object group a** or the **object group b**.

Deep-sky colour

From the menu: **Setup → Display → Deep-sky colour**

There are many different kinds of deep sky objects that you can display in your chart. With this dialog, you can give every sort of object their own colour, for a faster recognition on the chart.

You can choose preset configurations by selecting colour scheme, or make your own settings. To change colours, click the rectangled box, then select your colour from the pop-up window. You also can change the proportionality of the displayed surface brightness.

To configure your deep sky catalogs, go to the tab **CdC Deep Sky** from the menu **Setup → Catalog**. You can switch the display of deep sky objects on or off by pressening the ⚫ icon of the **object bar**.

Sky Colour

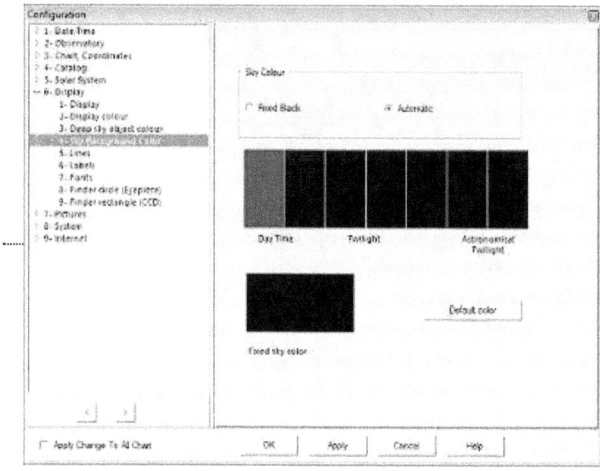

From the menu: **Setup → Display → Sky Colour**

With this dialog box you can choose the background colour for your chart. When the **Fixed Black** radio button from the *sky color* groupbox is selected, the background of the chart will allways have the same color. When you selected **Automatic**, the background-colour will respond according to the time: the colour chosen for the nautical and astronomical twilight, the day time, or when the Moon is above the horizon.

You can modify the fixed colour or twilight colour-schemes when you click on the colour areas.

You can switch between the fixed background colour and the configured colour scheme by clicking the ◨ icon from the <u>**marks group**</u> in the object bar.

Lines

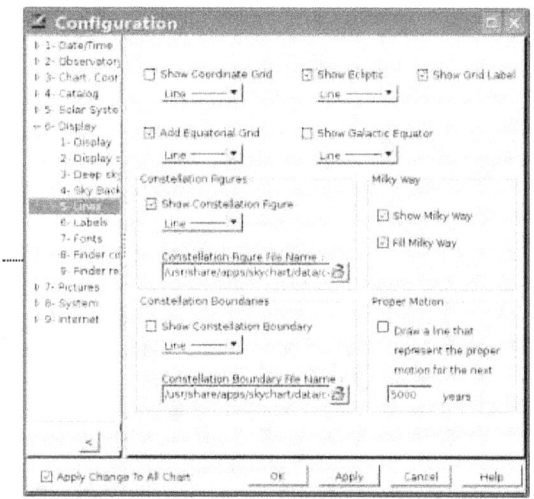

From the menu: **Setup → Display → Lines**

Here you can select which lines must be shown at the chart. Also you can choose in which line-mode they are drawn:

Show Coordinate Grid will draw the grid of the coordinate which is set in the menu **Setup → <u>Chart, coordinates</u>**. You can add the equatorial grid by marking **Add Equatorial Grid**. This grid will be very useful if your telescope has an equatorial mount. By checking the **Show Grid Label**, you will display the coordinates together with the grid on the chart.
If you want to display the ecliptic (the virtual line at the sky through which the planets, Moon and Sun seem to move for an observer on Earth), just click the **Show ecliptic** checkbox. And similarly, do you want to show the galactic equator, check the **Show Galactic Equator** checkbox.

Now you've got some more possibilities to make adjustments, in four groupboxes:

- **Constellation Figures**: If **Show Constellation Figure** is marked, CDC will draw the constellation lines.
- **Milky Way**: The checkbox **Show Milky Way** will activate the drawing of the boundary of the Milky Way. **Fill Milky Way** will fill it smoothly.
- **Constellation Boundaries**: Just like it says, **Show Constellation Boundary** will activate the drawing of the

boundaries of the constellations.

- **Proper Motion**: Every star has a proper motion in the Milky way. Here you can activate the display of the proper motion of the stars over the next xxx years.

An easy way to switch the display of these lines on or off, is by the **object bar** or by the menu **Chart → Lines/Grid**

Labels

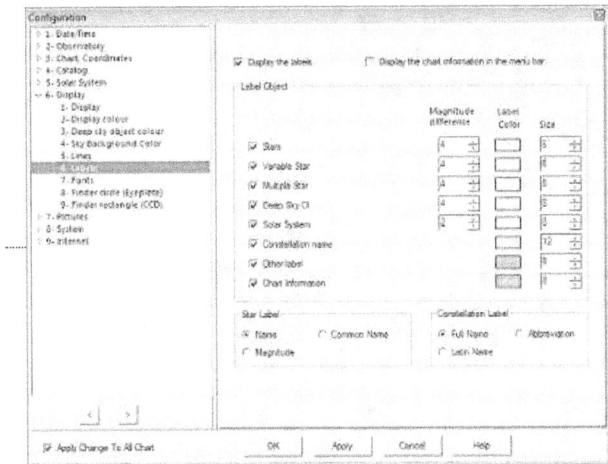

From the menu: **Setup → Display → Labels**

The **Display labels** checkbox controls the automatic display of all the labels, except the grid labels (controled by **Show Grid Label** in the **Lines** tab).

If you want to display chart information in the menu bar, check the **Display the chart ...** checkbox.

With the **Label Object** panel, you can control the colour and size of the font used in the labels according to the class of objects. **Other label** shows the compass directions on the horizon line.

In the **Star Label** and the **Constellation Label** panels you can specify the labels content.

Magnitude difference is a range (0 to 10) to filter labels according to the object magnitude: The lower this value, the more labels are displayed. The exact relation to configure labels to be displayed: the magnitude value of the displayed objects are smaller than **FOV number limiting magnitude - label Magnitude difference**. Example: FOV number limit: 6, Label magnitude difference: 2. **6 - 2 = 4**. As a result, objects with a magnitude value of smaller than 4 will be displayed with their labels, objects from magnitude 4 to 6 will be displayed without labels.

Labels of only one catalog at a time are displayed. The order of precedence is top-own, as is listed in the configuration of **star catalogs** and the **deep sky catalogs**.
You can change the font for you labels by **Setup → Display → Fonts**.

You can read more **here** about modification of individual labels, or about adding your own labels **here**.

Fonts

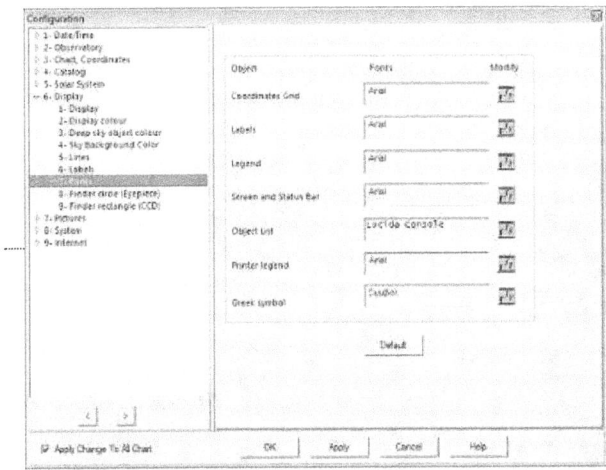

From the menu: **Setup → Display → Fonts**

By this tab, you can configure which font and its size to use in SkyCharts. To change a font, click on the **font** icon, at the right of the category.

Click on the **Default** button to reset your all your font settings.

Finder circle (Eyepiece)

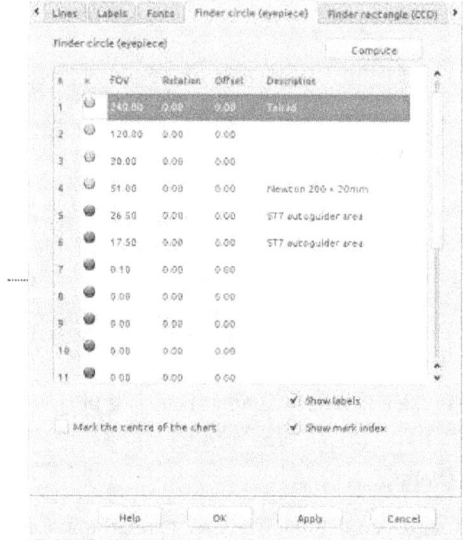

From the menu: **Setup → Display → Finder circle (Eyepiece)**

In this tab you can define the Field Of Vision (FOV) of your eyepieces, the unit is in minutes. By default, the eyepieces will be centered on the chart. Maybe you need the eyepieces to be off-axis, for example for guiding purposes. You can achieve this by configuring your eyepiece to be displayed with an offset, relative to the center of the chart (and centered eyepieces). Then it can also make sense to rotate your eyepiece.

Rotation is the angle (in degrees of arc) relative to the equatorial north, increasing order east, south west. You can change the rotation angle later when the circle is show on the map with the arrow keys.
The rotation angle can also be changed interactively from the chart display. Activate the rotation with the keys Shift+C to rotate the main camera, Shift+G to rotate the guider or Shift+S to rotate both at the same time.

Offset is a distance (in minutes) to the north, relative to the centered position. The offset doesn't read negative values, use an angle between 90 to 270 degrees to shift the FOV of your eyepiece to the south.

You can click the **Compute** button to open simple FOV calculator for your telescope and eyepiece.

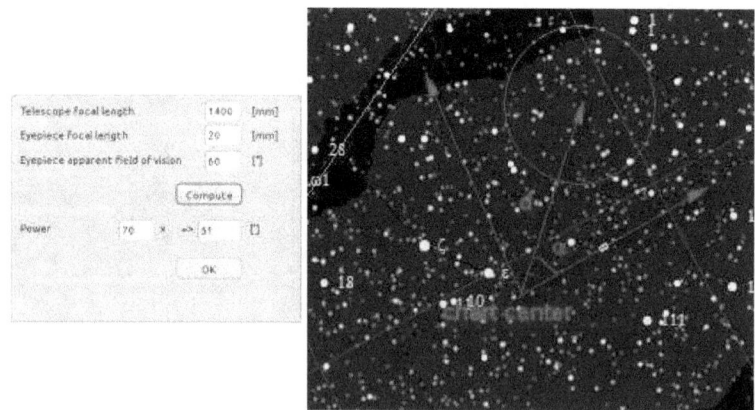

An easy way to switch the display of the eyepiece on or off, is by the ⊙ icon on the **object bar** or by the menu **Chart → Lines / Grid → Show Mark**.

Finder rectangle (CCD)

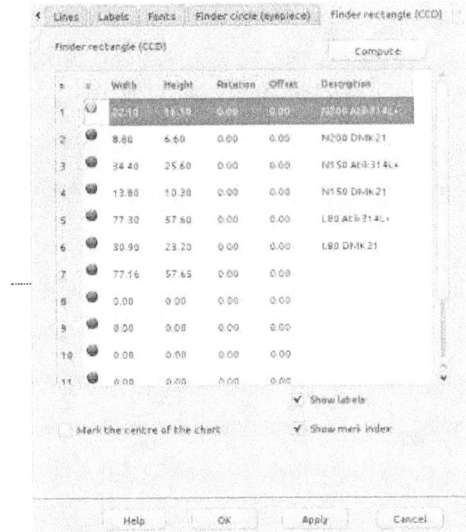

From the menu: **Setup → Display → Finder rectangle (CCD)**

Analog to the eyepiece settings, this tab allows you to define the FOV of your CCD cameras. Now, the field will be a rectangle with width and height in minutes. By default, it will be parallel to celestial equator, and centered on the chart. **Rotation** and **Offset** have the same meaning as with the eyepiece settings.

The rotation angle can also be changed interactively from the chart display. Activate the rotation with the keys Shift+C to rotate the main camera, Shift+G to rotate the guider or Shift+S to rotate both at the same time.

You can click the **Compute** button to open simple FOV calculator for your telescope and camera.

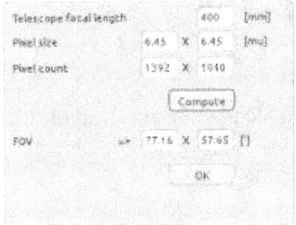

An easy way to switch the display of the eyepiece on or off, is by the ⊙ icon on the **object bar** or by the menu **Chart → Lines / Grid → Show Mark**.

Pictures

From the menu: **Setup → Pictures**

There are two locations to manage the FITS-formatted pictures:

- **install directory\data\pictures\sac** which contain the SAC object pictures.
- **user data\pictures** which is used to store the downloaded pictures from the DSS or the temporary pictures from RealSky.

DSO Catalog pictures

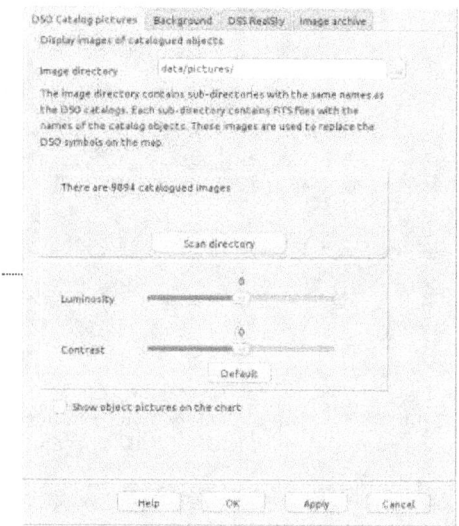

From the menu: **Setup → Pictures → Objects**

SkyChart can display deep sky objects in your charts in a more realistic way by showing FITS pictures. This can be independent of the display of symbols for deep sky objects. It can do this in any field of vision.

Before SkyChart can do so, the program needs to categorize all those pictures and load their characteristics into a database. It does this by a scan of the directory were you put those pictures, usually this is the **data/pictures** sub-directory of your SkyChart installation.

If necessary, you can change the path by entering it directly in the input area, or setting it with a dialog box after clicking the icon.
Please note that you must not specify the directory that contain the fits files itself but the directory just in front. This is because more package for other catalog may be available in separate directory in the future.

When this scan is finished, you will retrieve a message like: There are xxxx catalogued images (9894 in V3 beta 0.1.4).
If it stay with 0 image re-read the paragraph just ahead.

Don't forget to check "Show object pictures on the chart", at the bottom of the window. Otherwise these pictures will not be displayed on the chart. Also here you can set the Luminosity and Contrast of the images.

If you haven't installed any of those pictures, I can recommend you to install the SAC **picture package [http://www.ap-i.net/skychart/en/download]**.

Background Picture

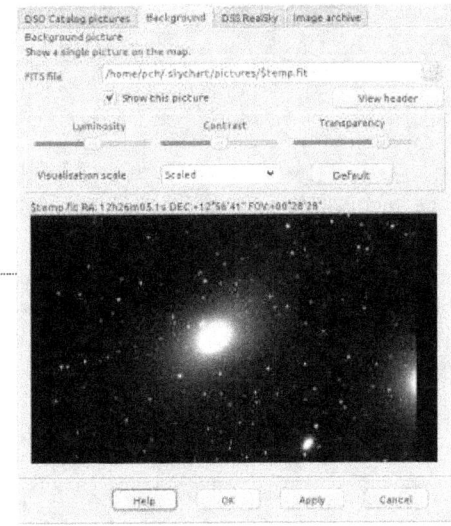

From the menu: **Setup → Pictures → Background**

Here you can open a FITS-file to display by entering its file name. After a new installation, the default directory to read from is **C:\Documents and Settings\[user]\Local Settings\Application Data\SkyChart\pictures** (Windows) or **/home/[user]/.skychart/pictures** (Linux). In this directory you usually will find the last FITS-file you downloaded from the DSS resources. (named "$temp.fit"). You can change the directory to any other source of FITS-files. For example, if you did install the SAC **picture package [http://www.ap-i.net/skychart/en/download]** with a typical installation of SkyChart, you can open FITS images of deep sky objects in your chart from the **/usr/share/apps/skychart/pictures/sac** (Linux) or **C:\Program Files\ciel\pictures\sac** (Windows) directory or After opening the file. The active chart will be repositioned according to the settings of the selected picture.

You can set the Visualisation scale, the Luminosity, the Contrast and Transparency of the image and you need to check "Show this picture".

DSS RealSky

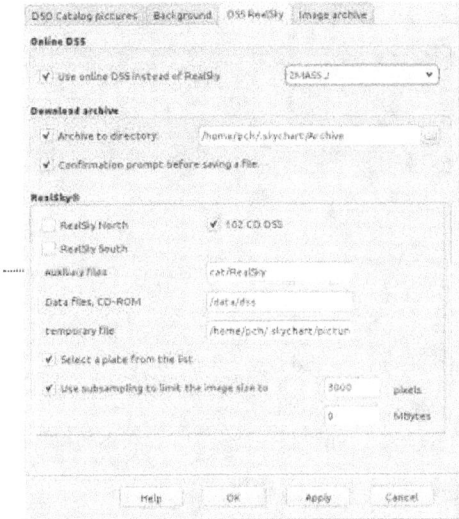

From the menu: **Setup → Pictures → DSS RealSky**

SkyChart can display FITS-picures for any place in the chart. This window is meant to configure from which sources to retrieve those pictures. You can get them by Internet, or if you have a RealSky CD-ROM set, from your CD-ROM player

Online DSS

If you are connected to Internet, you can download FITS-pictures from the Digital Sky Survey (DSS) site. With the combo box you can specify which set of DSS pictures you want to retrieve.

The size of pictures that you request are limited by the field of vision. Usually, you can't retrieve pictures when you set SkyChart to a FOV larger than two degrees of arc. The larger the FOV, the greater chances are that your request times out, or isn't supported by the server at all. You need to realize that it takes a lot of CPU power at the server site to generate a picture, you have to be patient.

To read more about how to download a picture from the DSS, click **here**.
To read more about the display of these images, click **here**.
To read more about the configuration of the DSS resources to download those images, press **here**.

Download archive

Use this section to archive on your computer the DSS picture you download from the Internet.
You can define here the directory to use for the archive and whether or not you want a confirmation prompt before to archive a file.
See next tab for the options how to display this archived images.

RealSky

Here you can set the RealSky CD-ROM location for your computer. You have to specify which set of CDs you are using, where the auxiliary files are to be found, the path to your CD reader and where the temporary files can be stored.
You can specify if you want to select a plate from a list. You can also choose to limit the size of the picture.

Image archive

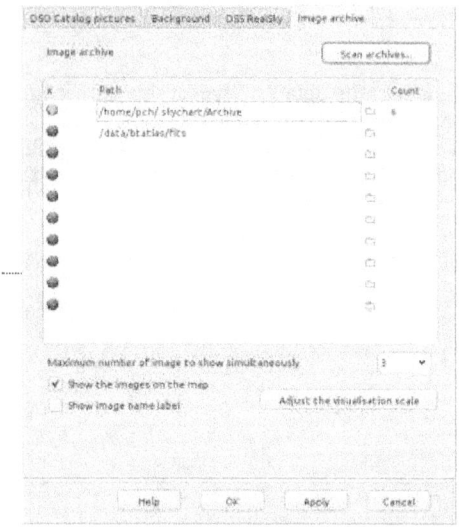

You can select here how to display the archived images.
Only the images in the FITS [http://fits.gsfc.nasa.gov/fits_home.html] format that include the WCS [http://fits.gsfc.nasa.gov/fits_wcs.html] information are eligible for use here. There is many software that allow you to run an astrometric reduction on your images, and also online [http://nova.astrometry.net] solutions.

The list can include up to ten directory where the program will search for picture to display on the map.

Select in the first column if you want this directory to be active now.

The second column is the directory itself. Be sure to always include the directory you select for the automatic archive of downloaded DSS images.

The last column indicate the number of pictures actually know in the program database. If you added some images manually in a directory you need to click the "Scan archives" button on top to update the database with the new files.

You can select if you want to show the images on the chart and if you want a label with the image file name.
You can also select the maximum number of images the program will show on a single map. Beware of performance issue if you select a too high value.
You can use the Image list window from a right click on the chart to change this setting for a single map display and also to select other images than the one automatically selected by the program.

Image list

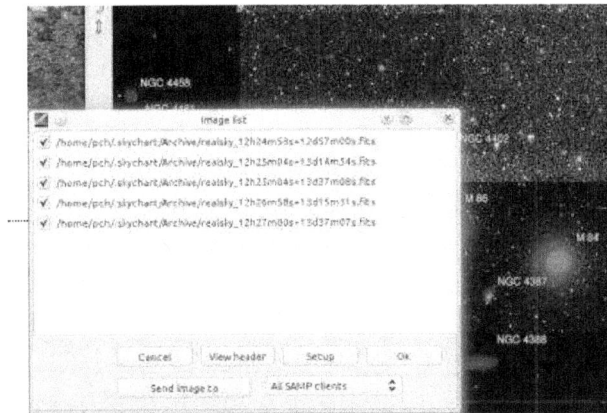

Use the chart right click menu to access this window.

The Image list let you see and alter which pictures are currently selected on the chart.
It list the last downloaded DSS image, or images from your Archives.

You can select which images to display by a selection in the first column. This selection is valid only as long the chart position and FOV is not changed. Click OK after you make your selection.

You can also view the FITS header for the selected file or open the Archives setup.

If you are connected to a SAMP hub you can send the selected image to another SAMP application.

General

General

In this window, you can manage the SkyChart database. It contains the orbital elements of comets and asteroids, SAC picture information, information about the DSS/RealSky pictures and information of observatory locations.
In order to display comets and asteroids, SkyChart needs to calculate the ephemeris, and store them in a database. But before SkyChart can do that, you need to download fresh files with **orbital elements** [http://en.wikipedia.org/wiki/Orbital_elements] for comets and asteroids. See the **comet** and **asteroid** tabs from the **Setup → Solar System** dialog box.

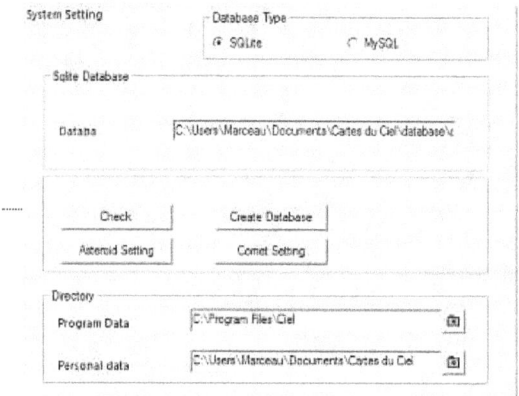

You may also want to create the SAC pictures information database. For more about installing and downloading SAC pictures, see the **Object** tab from the **Configuration → Pictures** dialog box.
The information about countries and the detailed information of Observatory locations are also stored in a database. You can create them by **Configuration → Observatory**.

A standard SkyChart installation creates a SQLite database. SQLite is very fast and for everyone simple to use. Use SQLite. Maybe you have reason to choose MySQL as DBMS. For example, maybe you wish to search comet and asteroid data external from Skychart, with other applications. Or you want to share one single database with many computers. You need to have at least basic knowledge about MySQL to do your queries. Another warning: When you decide to load the data from MPCORB without any limitation into your MySQL DBMS, it can take some hours. If you use Windows XP, copy the file libmySQL.dll from the bin directory of your MySQL installation to the installation directory of Skychart.

- **SQLite**: an input area contains the path to SQLite DB. With Windows XP, it is stored in \My documents\[user]\Local Settings\Skychart\database\. You can specify an alternate location.

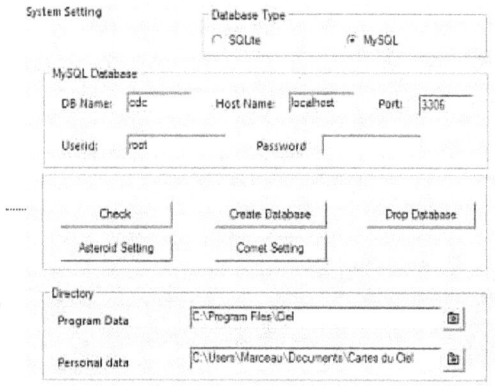

- **MySQL**: You need a MySQL server to connect to. A local server connection is proposed, which works fine with a MySQL local server installation. You can specify an alternate connection with default parameters:
  - **DB Name** cdc
  - **Host Name** localhost
  - **Userid** root
  - **Password** [your mysql root passwd]
  - **Port** 3306 (standard MySQL port)
  - **Asteroid Setting** directs yous to **Configuration → Solar System → Asteroid**. (older SkyChart versions)
  - **Comet Setting** directs you to **Configuration → Solar System → Comet**. (older SkyChart versions)

- **Drop DataBase** this button deletes all database content.

- **Create DataBase** button launches the SQL "define tables" script.
- **Check** lets you verify if the tables succesfully are created.

Directory group specifies the SkyChart installation path and personal (Your_Documents\Carte du Ciel\) data path. You can specify alternate locations for a non standard installation. Help yourself with the directory icon at right of the input area.

Server

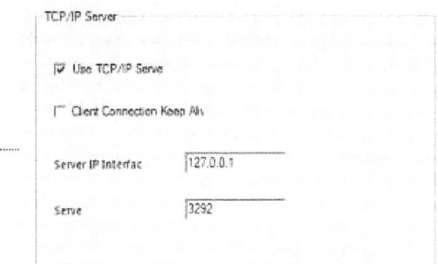

Here you can set connection parameters to use Skychart as a server. You can verify the connection states in **View → Server information**

- SkyChart accepts connections by checking **Use TCP/IP Server**.
- SkyChart verifies the attendance of the client and closes the connection if the client is not active by checking **Client Connection Keep Alive**.
- **Server IP interface** define the network interface Skychart is listening on. The default 127.0.0.1 allow only for local connection. If the client is remote, you need to set your interface IP address or 0.0.0.0 to listen on all the interfaces.
- SkyChart is listening on the port set in **Server IP Port**, if you change this, you must change client software correspondingly.

Telescope

From the menu: **Setup → System → Telescope**

Before you can use SkyChart to work with your telescope, you need to do some basic settings here. First, select which interface driver you will use. Depending on your choice, the content of the dialog box changes. You can choose from:

ASCOM

Use this on Windows only. With ASCOM you can drive most astronomical equipment like your mount and dome. If you don't have the Ascom driver installed, you can download it from http://ascom-standards.org/ [http://ascom-standards.org/]

Indi driver

Use this on Linux or Mac only. The Indi package [http://www.indilib.org] is designed to use with all sorts of astronomical devices. Many computer operated mounts and domes can be driven by Indi. You can **download [http://www.indilib.org/download.html]** this driver if it didn't come with your Linux distribution. When you want to use the Indi driver, make your specific settings in the dialog.

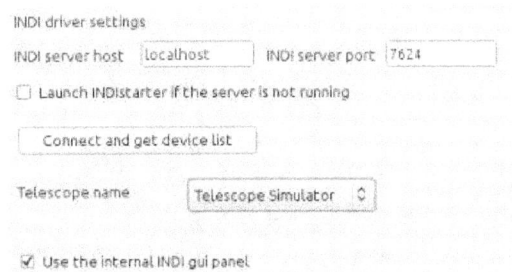

After setting the server host and port, click "Connect and get device list" to fill the list of driver supported by your INDI server. Then select the telescope driver from the list.

It may be difficult to know the device serial port name if you have many adapter or if you use a USB-serial adapter. In this case use the dmesg command to find this name.
In the following example you must indicate /dev/ttyS0 for the motherboard port or /dev/ttyUSB0 for the USB port.

```
serial8250: ttyS0 at I/O 0x3f8 (irq = 4) is a 16550A
```

LX200

This driver supports LX200, Autostar, Magellan I/II and other systems using the LX200 protocol. When you are using **Mel Bartels' stepper motor system [http://www.bbastrodesigns.com/BBAstroDesigns.html#Computer_Operated_Telescopes]** to drive your mount, you can use this driver.

On the main bar, you will find the **telescope group** icons. After you made the proper basic settings for your computer operated telescopes, you need to configure the your driver by a click on the ⚙ icon to drive your telescope in the right way. You need the ⊕ and 🔭 icons from the main bar to synchronize your mount with an object on the chart or to slew it to another object. See for more information about this the menu **Telescope**.

Encoder

You can use this with an encoder box using Tangent protocols like the Ouranos, AAM or NGC-MAX.

Manual Mount

When you use this, SkyChart asks you to give the settings for your specific mount. Choose your mount type from **Equatorial Mount** or **Alt/Az Mount**. Next, you set how many full turns you must make on your knob to slew your mount by one degree of arc. Usually, manual mounts are wormgear driven. Most common is a mount with 144 tooths on the worm wheel for both directions. When you know the number of tooths, calculation of the number of turns per degree of arc, or per hour is easy:
Per degree: 144 / 360 = 0.4
Per hour: 144 / 24 = 6
Also you need the find out how to make the direction setttings for the knobs. Only when a **counter-clockwise** turn of a knob causes an **increase** in the right ascension, azimuth, declination, or altitude, you need to check the appropriate `Reverse .. Direction`. Otherwise, leave it unchecked. From now on, you can find instructions how to turn the knobs of your mount to move from one object to another. To find these, first click the object on the chart to which your telescope is pointed. Next, click on the object you want to observe. Now click on the label to retrieve the information details. The last lines show you who to turn the knobs on your mount.

Language

Language selection

en - English

Simply select the language from the combo box that you want SkyChart to use.

If your language isn't there, why not make your own GUI translation? It 's not hard, only takes a few hours. Have look **here [http://www.ap-i.net/skychart/en/translations#software_translation]**.

SAMP

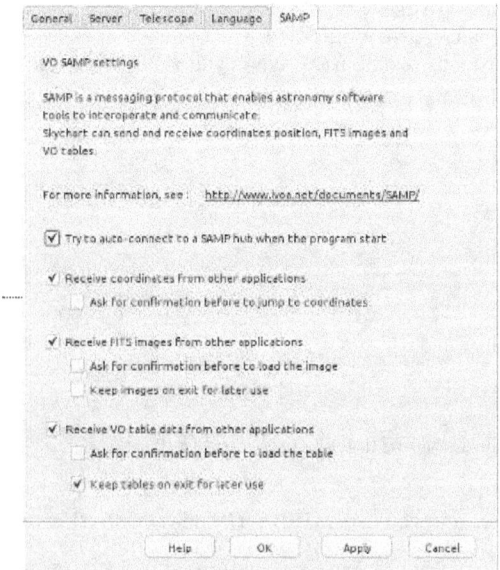

Set the default options for the SAMP interface.

- **Try to auto-connect** : Connect automatically to a running SAMP hub when the program start.
- **Receive coordinates** : Allow the program to jump to coordinates received from other applications. For example a click in Aladin.
- **Receive FITS images** : Allow the program to receive an image from other applications. From Aladin use the menu "Interop / Broadcast selected image to…"
- **Receive VO table data** : Allow the program to receive data table from other applications. From Aladin use the menu "Interop / Broadcast selected table to…".

You can request a confirmation before to accept any data and keep the data you receive for offline use. The default is to cleanup the data when the SAMP connection is closed.

It is also possible to send data from Skychart to the other applications.

- To connect to a hub after the program is started, or to get the connection status, use the menu File -> SAMP.
- To send coordinates use the right click menu SAMP send coordinates
- You can send table data from the VO catalog setup.
- To send a FITS image use the right click menu Image list, the "Send image" button.

ASCOM Telescope interface

From the menu: **Telescope**

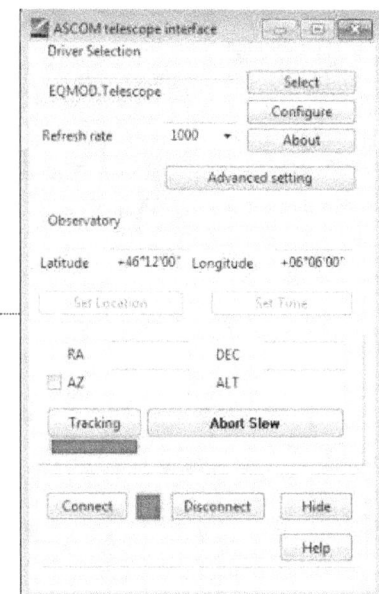

This interface can be used with any telescope supported by the ASCOM Platform. It work only on Windows. Please refer to http://ascom-standards.org [http://ascom-standards.org] to obtain the latest driver version and for more information.

Then use Setup → System → Telescope to select the Ascom interface and then open the Control panel.

For the first time use you need to **Select** the driver to use and provide some configuration information depending on the telescope you use.

The **Advanced setting** button open a window with some settings you normally not need to change.
At the moment this is use to force the equatorial system equinox when it is wrongly reported by a driver.

Then click the "Connect button", the light must change to green and the telescope coordinates are displayed.

You can now set the geographical coordinates and time of the telescope using the values from your computer. But this is normally not required as your telescope must be already aligned before you connect.

To free some space on the screen you can close the interface window by clicking the "Hide" button. Refer to the main menu Telescope to learn how to show the telescope position on the chart or to use the Goto facility if supported by your telescope.

Also if supported you can refine the local slewing precision by using "Sync current object" on a nearby star at any time.

The configuration options are saved when you hide the window. Also save the default options in Skychart menu to keep your interface choice.

INDI Telescope interface

From the menu: **Telescope**

The configuration of the telescope driver is done from the menu <u>Setup -> General -> Telescope</u>

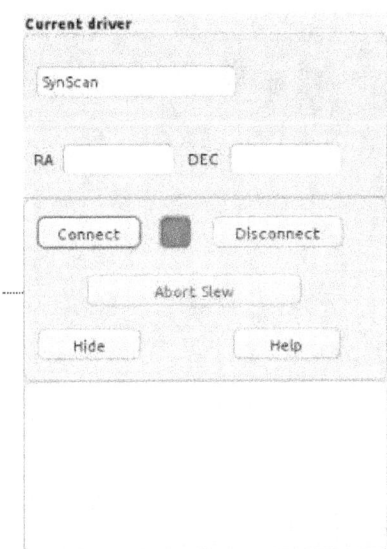

Click the "Connect button", the light must change to green and the telescope coordinates are displayed.
In case of connection problem you can look at the message at the bottom of the window.

To free some space on the screen you can close the interface window by clicking the "Hide" button. Refer to the main menu <u>Telescope</u> to learn how to show the telescope position on the chart or to use the Goto facility if supported by your telescope.

Also if supported you can refine the local slewing precision by using "Sync current object" on a nearby star at any time.

LX200 Telescope interface

From the menu: **Telescope**

Telescope connection

This interface can be used with LX200, Autostar, Magellan I/II and other systems using the LX200 protocol.

Please refer to the manufacturer information to connect the telescope to the computer and power on the telescope. Also perform now the initialization procedure for your telescope, this must be done before to connect to the program.

Then use Setup → General → Telescope to select the LX200 interface and open the Control panel. The following window is displayed :

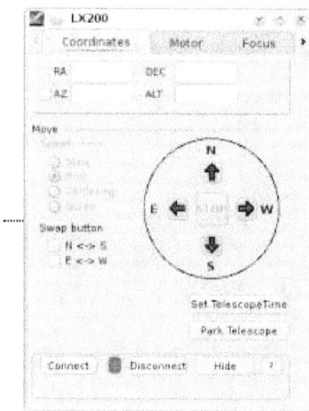

For the first time use you need to provide some configuration information, see below.

Click the "Connect button", the light must change to green and the telescope coordinates are displayed at the top of the screen.

You can now close the interface window by clicking the "Hide" button. Refer to the main menu Telescope to learn how to show the telescope position on the chart or to use the Goto facility.

Remember you can refine the local precision by using "Sync current object" on a nearby star at any time.

You can also use the virtual pad to move the telescope around, be careful the select the right model in the configuration before to use this function.

If you use a LX200 with the high precision pointing mode the button to continue the slew is also here.

Interface configuration

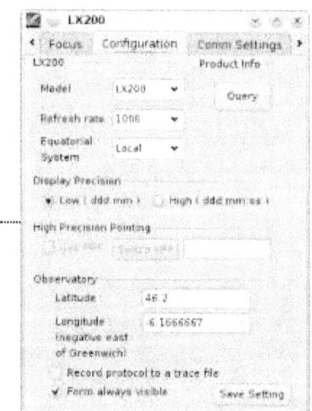

Select the telescope model you use, if your model is not listed consult your documentation to find a compatible model, if you find none select the LX200 model if slewing is supported or Magellan II if not.

The Refresh Rate is the amount of time (in millisecond) that elapse between two position query to the interface. Use a smaller value for a smoother cursor movement if your telescope accept that and if your computer as enough power. Use a larger value to use less computer resource or if your telescope as a limited output rate.

Select the precision used to transfer the data between the telescope and the program, this may help to solve some communication problem. Select if you want to use the LX200 High Precision Pointing, this can be checked only after you connect to the telescope.

You don't need to set the observatory coordinates as this is automatically done from the location defined in Skychart.

In the case of problem with the interface you can trace the protocol to a file by checking "Record protocol to a trace file".

The last check box let you choice if you want this window always visible at the top of the other.

Communication port configuration

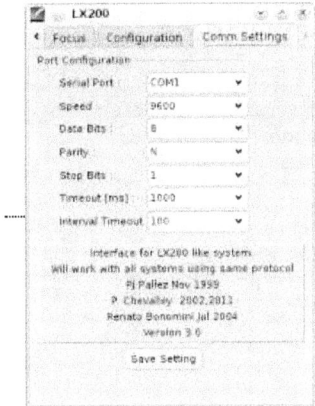

Select the serial port where the telescope is connected (COM1,... on Windows; /dev/ttyS0 or /dev/ttyUSB0 on Linux; /dev/tty.serial1 or /dev/tty.usbserial-xx on Mac).

Set other options (speed, data, parity and stop) according to the manufacturer informations. Generally LX200, Autostar and Magelan II use 9600,8,N,1 but you may change to 1200, 8, N,1 and Timeout: 500ms for the Magelan I.

Change the timeout values only if you encounter communication problem and after identifying the problem with the trace file. Do not set the timeout to a value greater than the refresh rate in the preceding screen.

Click the Save Setting button to keep your configuration for the next run. Also save the default options in Skychart menu to keep your interface selection.

Encoder Telescope interface

From the menu: **Telescope**

Telescope connection

This interface can be used with an encoder box using Tangent protocols like the Ouranos , AAM or NGC-MAX. It uses a floating two stars alignment method based on an article by Toshimi Taki [http://www.geocities.jp/toshimi_taki/aim/aim.htm] in February 1989 S&T.

Please refer to the manufacturer information to connect the telescope to the computer and to power. Then use Setup → General → Telescope to select the encoder interface and open the control panel. The following window is displayed:

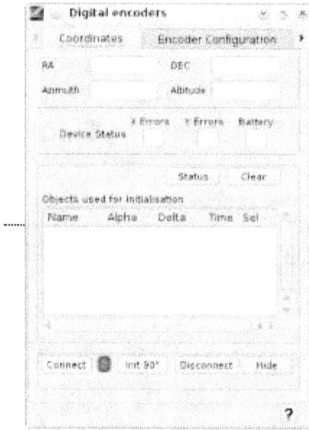

The first time you use this feature, you need to provide some configuration information.

Click the "Connect button", the light must change to green and the encoder count is displayed at the top of the screen.

Perform the following alignment procedure:

- If you select 90 degree initialization:
    1. Point the telescope vertically for an altaz mount or parallel to the polar axis for an equatorial mount. It is important to understand that you must not point to the real zenith or pole but to the direction of your mount axis. Use of a fixed graduated circle or a stop piece can be of great help; see also the discussion about the Z3 parameter below.
    2. Click the "Init 90°" button.
- If you select 0 degree initialization:
    1. Point the telescope horizontally for an altaz mount or perpendicular to the polar axis for an equatorial mount. The above note also applies in this case.
    2. Click the "Init 0°" button.

- Point the telescope to a star, use a reticle eyepiece and enough power to precisely center the star.
- Click the star on the chart.
- Select "Sync current object" from the "Telescope" menu or the right mouse button menu.
- Repeat the procedure for a second star. Be sure to choose a star with enough coordinates difference on each axis.

The position of the telescope is now displayed using equatorial and azimuthal coordinates.

You can now close the interface window by clicking the "Hide" button.

Refer to the main menu Telescope to learn how to display the telescope position on the chart.

Remember you can refine the local precision by using "Sync current object" on a nearby star at any time. You are not limited in the number of initialization stars you choose.

The couple of stars used for the part of the sky the telescope is actually pointed at are marked by a '*'.

The Clear button deletes all the stars in the initialization list and returns the interface to the uninitialized mode.

A right mouse click on the list allows the deletion of one star from the list.

If your interface returns status information, this is indicated just below the coordinates.

Main causes of errors

Sometimes the position shown on the chart is not the one you observe in the telescope. There are many causes of possible error - here is a short list:

- Configuration error:
  - Number of encoder steps incorrectly set, principally if you gear the encoder. Use a full mount rotation with a fixed remote reference to count the exact number.
  - Mount fabrication error not set.
  - Mismatch between azimuthal and equatorial mount.
- The precision for two star alignment is dependent on a precise first direction initialization, use the Z3 parameter to correct a systematic error. There is, however, no need for a precise polar alignment or mount leveling.
- Unadapted reference stars, the first two alignment stars must have enough difference along the two axes and be far enough in the sky.
- Also avoid selecting an alignment star near the direction of the polar axis; Polaris is good for an azimuthal mount but not for an equatorial mount.
- Missing encoder count in the interface box, this may occur if you move the mount too rapidly or if the battery is low.
- Misidentification of a reference star.

Interface configuration

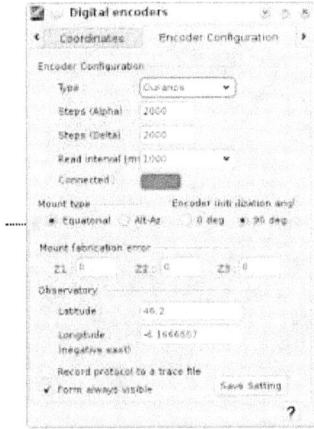

Select the encoder type you use. If your model is not listed consult your documentation to find a compatible model, if you find none select the generic Tangent model.

Set the encoder count for a full rotation for both axes.

The read interval is the amount of time that elapses between two position queries to the interface box. Use a smaller value for a smoother cursor movement if your box accepts that and if your computer as enough power. Use a larger value to use less computer resource or if your interface box has a limited output rate.

Check the mount type (equatorial or Alt-Az) you use and if you prefer to do the first initialization vertically or horizontally.

The mount fabrication error angle Z1, Z2, Z3 in degrees are the ones defined in Taki article:

- Z1 is the amount by which the elevation axis is offset from the perpendicular to the horizontal axis.
- Z2 is the pointing error of the telescope optical axis in the same plane.
- Z3 can be considered a correction to the zero setting of the elevation circle.

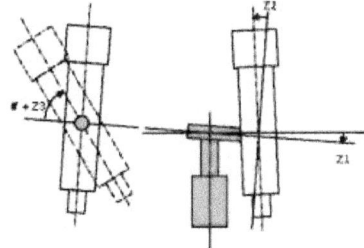

Measuring Z1 and Z2 can be a difficult task, but Z3 can be of great value in initializing the mount.
For example, if your Dobson mount as a security stop 5 degrees above the horizon to avoid flipping the primary mirror. You can set Z3=5 and check the initialization at 0 degrees. To initialize the mount simply point the telescope tube horizontally to the security stop and click the "Init 0°" button, that's all.
You can do the same if the stop is 15 degrees after the zenith but in this case use the "Init 90°" button.

You don't need to set the observatory coordinates as this is automatically done from the location defined in Skychart.

If there is a problem with the interface you can trace the protocol to a file by checking "Record protocol to a trace file".

The last check box lets you choose if you want this window always visible at the top of the other.

Communication port configuration

Select the serial port where the telescope is connected (COM1,... on Windows; /dev/ttyS0 or /dev/ttyUSB0 on Linux; /dev/tty.serial1 or /dev/tty.usbserial-xx on Mac).

Set other options (speed, data, parity and stop) according to the manufacturer's information.

Change the timeout values only if you encounter communication problems and after identifying the problem with the trace file. Do not set the timeout to a value greater than the refresh rate in the preceding screen.

Click the Save Setting button to keep your configuration for the next run.

Also save the default options in Skychart menu to keep your interface choice.

Internet

Proxy

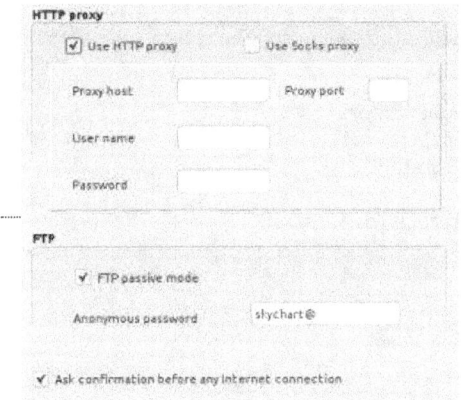

From the menu: **Setup → Internet → Proxy**

- **HTTP Proxy**

If you are connected to Internet through a proxy, SkyCharts needs its parameters to work.
The parameters are:

- **Proxy host** IP address or hostname of the proxy.
- **Port** port number used to communicate with the proxy.
- **User Name** Name used to connect to proxy.
- **Password** password used to connect to proxy.

The default is not to use a proxy. If you don't need a proxy, then you don't need those parameters.

- **FTP** is used by SkyChart to download the larges files from MPC, an anonymous connection is required. If you are nice, you put your email address as a password here. Anyhow, it usually will work with anything that contains an "@" in the password.

You can skip to push the download button every time you want to retrieve data from online resources. To do so, you need to uncheck the checkbox "Ask confirmation before any Internet connection" at the bottom of this tab.

Orbital Elements

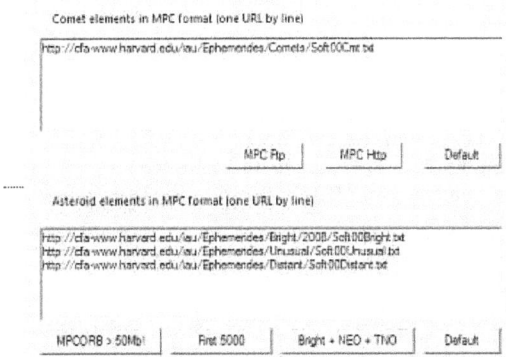

From the menu: **Setup → Internet → Orbital Elements**

Here the MPC URLs are specified that SkyChart can use to download the orbital elements of comets and asteroids.

Comets

The upper input area contains the URLs for comets, you can choose the site and protocol to download from by a click on one of the four buttons:

- **MPC Ftp** comet data from **cfa-ftp.harvard.edu [ftp://cfa-ftp.harvard.edu/pub/MPCORB/COMET.DAT]** by the FTP protocol
- **MPC Http** same as above, only **HTTP protocol [http://cfa-www.harvard.edu/iau/Ephemerides/Comets/Soft00Cmt.txt]**
- **Default** sets the URL to the same as MPC HTTP.

Asteroids

The lower input area shows the MCP URLs that are used for asteroids, you can choose the source for the orbital elements by clicking one of the six buttons:

- **MPCORB > 70MB!** for full file (more than 300.000 asteroids, FTP protocol mandatory)
- **First 5000** for a choice of 5000 asteroids (file downloaded from CdC Site)
- **Bright + NEO + TNO** For three MPC lists of asteroids (bright, unusual and distant)
- **Default** is same as **First 5000 + NEO + TNO**.
- **astro.cz > 70B!** A MPC-mirror situated in the Czech Republic. 300.000 asteroids, by HTTP
- **First 5000 + NEO + TNO** 5000 asteroids + the unusual and distant

Online DSS pictures

From the menu: **Setup → Internet → Online DSS**

Here the DSS URLs are specified that you can use to download FITS formatted pictures. Each row in the list has a name and a URL. The name is used as a short identifier for the long URL. The URLs describe where and exactly how to download the pictures from. (see for example the DSS Site [http://archive.eso.org/cms/catalogs-dss] for details).

The names are used to select the source of your pictures, make your selection by a click on the DSS icon in object tool bar, or by clicking **Chart → Get DSS Image**.

When you are an experienced user and you know pretty well what to do, you can add more rows with this dialog box. You will not break anything if the syntax is wrong, the worst that can happen is that the program will not be able to retrieve a picture by a faulty URL.
First you need to find out the syntax and the type of parameters that the website can handle. When you know this, you need to fill the variables for the parameters with the values that automatically will be delivered by Cartes du Ciel-Skychart.
Now, just click an empty line, enter a short name at the left. At the right, enter the complete URL with the proper variables from Cartes du Ciel-Skychart on the right places.

The next list gives you an overview of the variables that Cartes du Ciel-Skychart uses to generate the query URLs.

SkyChart variable	meaning
$RAH	Center coordinates, Right Ascension Hours
$RAM	Center coordinates, Right Ascension Minutes
$RAS	Center coordinates, Right Ascension Seconds
$DED	Center coordinates, Declination Degrees, North of celestial equator is prefixed with +, south prefixed by -

$DEM	Center coordinates, Declination Minutes
$DES	Center coordinates, Declination Seconds
$RAF	Center coordinates, Right Ascension, in decimal degrees of arc
$DEF	Center coordinates, Declination, decimal degrees. North of celestial equator is represented without a sign, south prefixed by -
$XSZ	FOV in X direction, decimal minutes of arc, precision 3 digits
$YSZ	FOV in Y direction, decimal minutes of arc, precision 3 digits
$FOVX	FOV in X direction, decimal degrees of arc, precision 6 digits
$FOVY	FOV in Y direction, decimal degrees of arc, precision 6 digits
$PIXX	Pixels in X direction
$PIXY	Pixels in Y direction

To read about the URL-format in queries, read the **ASU document [http://cdsweb.u-strasbg.fr/doc/asu.html]**.

To read more about how to download a picture from DSS, click here.
To read more about the display of these images, click here.
To read more about the configuration of the usage of DSS, click here.

Artificial satellites

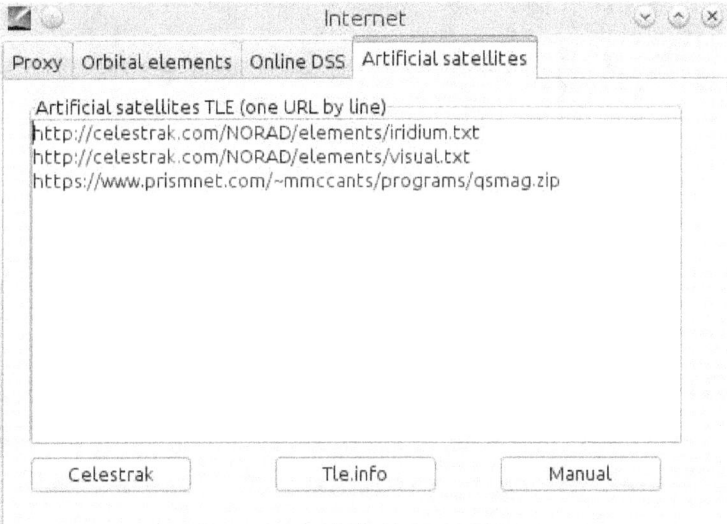

From the menu: **Setup → Internet → Artificial satellites**

Here the URL for the TLE files you want to download using the Calendar **Download TLE** button..

You can specify TLE files with an extension of .tle or .txt, or a .zip containing one of this files.

Click one of the button to fill the list with example URL.
You can add other files as you want, take a look at the source web site to get a list of the available files:
http://celestrak.com [http://celestrak.com]
http://www.tle.info [http://www.tle.info]

Keep the qsmag.zip line as it update the satellite magnitude estimate file.

Click the Manual button to clear the list and download manually the tle from https://www.space-track.org [https://www.space-track.org] as in the previous version.

Labels

The selection of the labels to be shown

You can configure the automatic display of labels in SkyChart by **Setup → Display → Labels**.
You also can set the font type, the font size, color, and which kind of objects to label.
For stars and constellations you can set the type of content to fill the labels.

The modification of labels

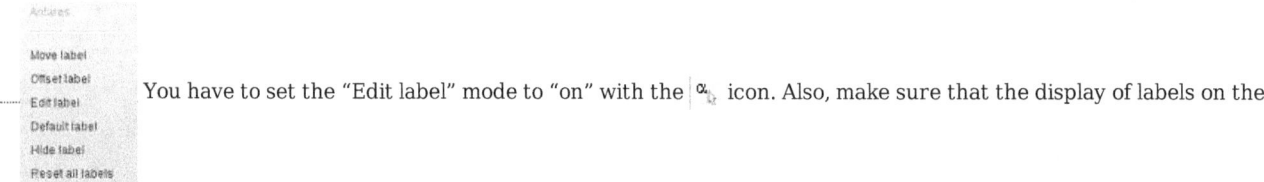 You have to set the "Edit label" mode to "on" with the α icon. Also, make sure that the display of labels on the

chart is enabled by checking that the α icon is pressed. Then, if you right-click on a label, a pop-up window will be diplayed

with these entries:

- **Move label** changes the mouse cursor to become a cross, by moving the mouse you can drag the label over the chart until you do a left mouse click in the chart.
- **Edit label** a window pops up which enables you to enter a new text as the label.
- **Default label** restores the label to its original content and its place.
- **Hide label** as it says.
- **Reset All Labels** undo all label modifications.

If you want to read more about the pop-up windows from the chart, click **here**.
To read more about the automatic display of labels, read **this**.

Add user labels

To add a label, first you need to provoke a pop-up window by a right-click on an object or anywhere on the chart. (If you want more information about the pop-up window, see **Pop-up windows**.)

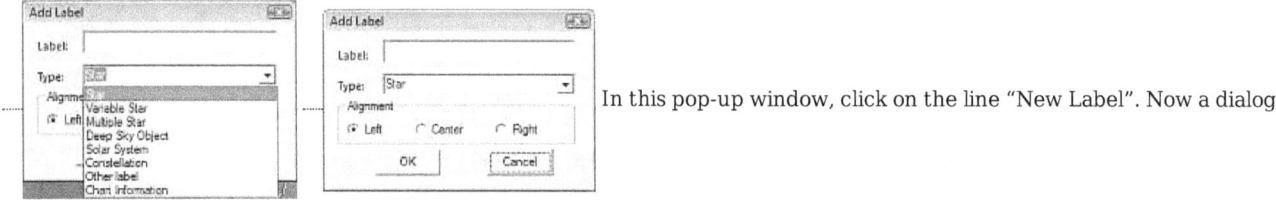 In this pop-up window, click on the line "New Label". Now a dialog

box will open with the next items:

- **Label** is an input area where you can enter the displayed text of the label.
- **Type** is a combo-box where you can choose from which type the labeled object is.
- **Alignment** is a radio-button group to position the label near the object.

Advanced Search

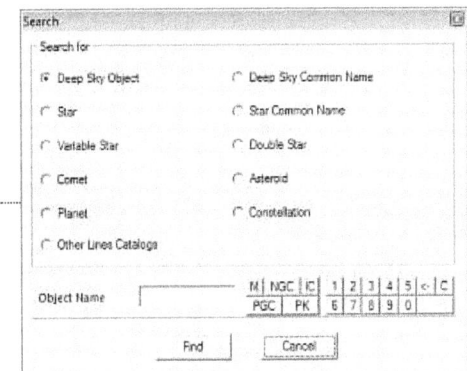

From the menu: **Edit → Advanced Search**, or from the icon on the **Search group** of the main bar.

First, select the appropriate object type by a click of one of the radio buttons. Then enter its ID in the bottom entry area.

The buttons can help you to enter the catalog - and the object id at the right hand side of this area. When the object is found, the chart will be shifted to display the object in the centre.

If you do not get a result, it may be because SkyChart could not find the object in the configured catalogs. Click **here** to read more about configuring catalogs, or click **here** to read about the installation of extra catalogs.

Position and Field of Vision

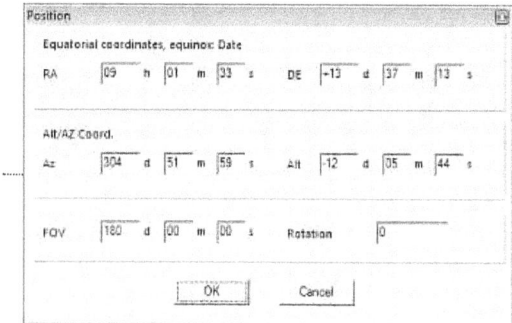

From the menu **View → Position** You can also open this window by clicking on ⚲ icon in the **main tool bar**.

Position

This provides a quick and easy way to read or set the coordinates of the center of the chart. Coordinates can be equatorial or azimuthal. The conversion is done automatically.

You also can read the equinox setting of the equatorial coordinate system that your chart uses. If you want to change this, you can specify the equinox (and epoch) for your chart by the tab Chart, Coordinates from the menu by **setup → Chart, Coordinates**, in the **type of coordinates** part.

You can also change the position in a less controlled way, see for that the **directions and displacements** shortcuts.

Field of Vision (FOV)

In the lower part of the position dialog box you can set a very precise FOV (precision: second) and a rotation (precision: degree) for the chart. Simply fill the size of your FOV or the rotation in.

Other ways to change the field of vision: You can change the FOV in an easier -but less precise- way by the **zoom group buttons** from the main bar. Or by the buttons from the **The Field of Vision group**. Or with the **shortcuts**.

Observing List

This is a simple observing list capability that let your prepare in advance a list of object to observe.
You can filter the visible objects using a few criteria, sort the list on any column, advance to the next or previous object.

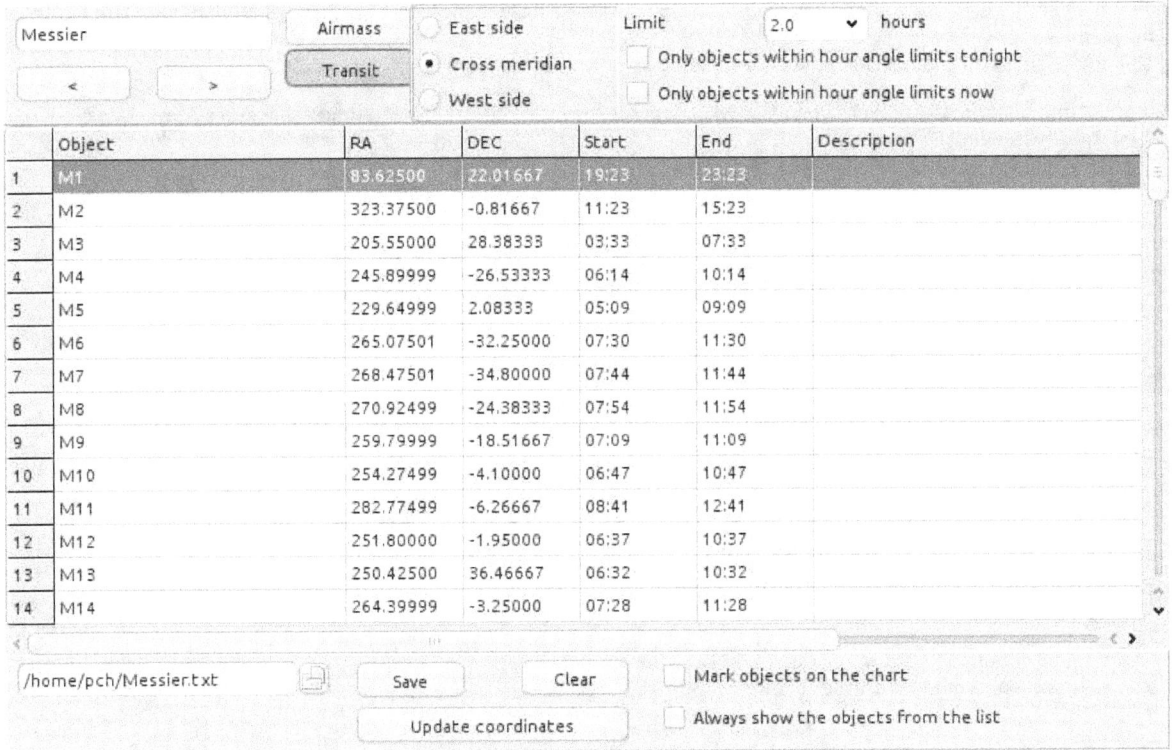

You can start with an empty list and add the object from the chart right click menu.

Or create an initial file with a title line followed by the object name on every next lines, like this Messier list [http://www.ap-i.net/pub/skychart/catalogues/messier.txt] example. Many planning software let you output such a file.
The coordinates are added when the file is first loaded.

Don't forget to Save the file after you make some change.
The Clear button allow to start a new empty file with the default name.
The Update coordinate button refresh the coordinates from the Skychart catalogs.

The object from the list can get a special label to distinguish them, check "Mark objects on the chart" for that.
The other checkbox "Always show the objects from the list" is to be use in special case when the objects are very faint and you want to show them on large scale chart. Beware this option can severely degrade the performance.

The RA and DEC column are the object coordinates as read from the catalogs. Both are show in degree units.

You can enter a free text in the Description column. For example the time of observation during your Messier marathon.

The Start and End Time are computed based on the filter selection.

You can make a selection based on the transit time.

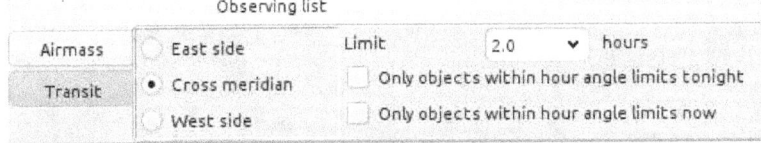

In this case indicate the side of the meridian (important to avoid a meridian flip with a German equatorial mount), or both, and the time limit or hour angle from the meridian.

Then select if you want to filter the list with all the object that will get through this criterion during the night, this is useful for planning.
Or filter only the object that fill the criterion right now, useful when you are in an observing session.

You can also make the selection based on the object minimal altitude, or maximal airmass.

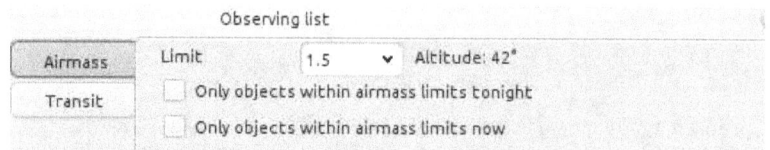

In this case select the maximal airmass your observation can afford or select "Horizon" if a simple detection near the horizon is acceptable. Then select if you want to filter the list with all the object that will get through this criterion during the night, this is useful for planning.

Or filter only the object that fill the criterion right now, useful when you are in an observing session.

Sort the list by End time to know the objects you need to observe in priority. Note that the time scale start from noon to ensure the list continuity during the night.

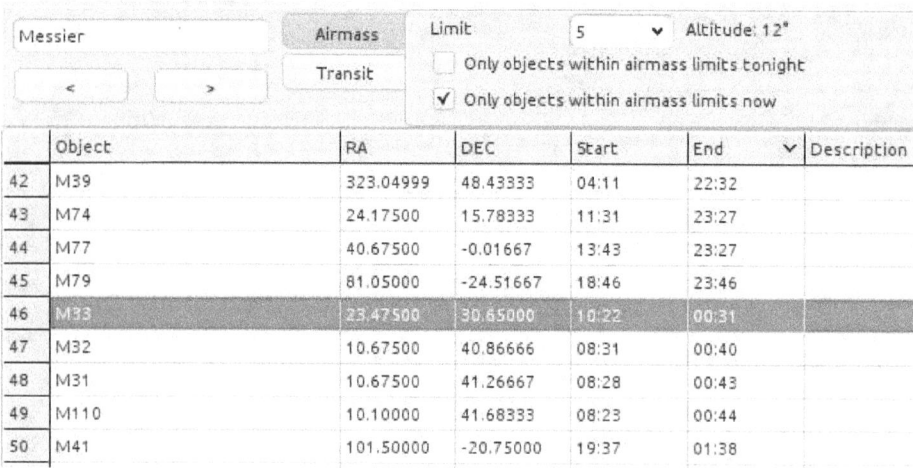

	Object	RA	DEC	Start	End	Description
42	M39	323.04999	48.43333	04:11	22:32	
43	M74	24.17500	15.78333	11:31	23:27	
44	M77	40.67500	-0.01667	13:43	23:27	
45	M79	81.05000	-24.51667	18:46	23:46	
46	M33	23.47500	30.65000	10:22	00:31	
47	M32	10.67500	40.86666	08:31	00:40	
48	M31	10.67500	41.26667	08:28	00:43	
49	M110	10.10000	41.68333	08:23	00:44	
50	M41	101.50000	-20.75000	19:37	01:38	

Click on the row number column to center the object on the chart.
A right click on a row present a menu with the following options:

- View on chart
- Update coordinates
- Delete

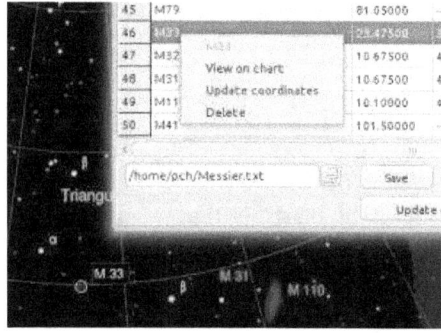

Click the Save button to save the content of the observing list.

To save the selection and other options save the main program configuration from the menu Setup → Save configuration now.

It is possible to automate the Observing list by using the Server Commands. An example script that point the telescope to each visible object in sequence is available here:
http://sourceforge.net/p/skychart/code/HEAD/tree/trunk/skychart/sample_client/python/
[http://sourceforge.net/p/skychart/code/HEAD/tree/trunk/skychart/sample_client/python/]

Virtual Observatory SAMP interface

SAMP [http://www.ivoa.net/samp/] is a messaging protocol, part of the Virtual Observatory, that enables astronomy software tools to interoperate and communicate. Skychart can connect to this other tools [http://wiki.ivoa.net/twiki/bin/view/IVOA/SampSoftware] and send or receive coordinates position, FITS images and VO tables or selections.

Very useful tools to use with Skychart are Topcat [http://www.star.bris.ac.uk/~mbt/topcat/] and Aladin [http://aladin.u-strasbg.fr/aladin.gml].

Skychart do not include a hub, so you need to connect to another software that include this piece.

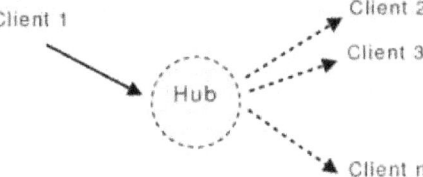

Initialization

To start to use SAMP with Skychart you can configure some options from the menu Setup -> General -> SAMP. This include how to connect to the hub, which function request to accept, and how to keep the data after the program is closed.

If you select to not connect automatically, or if the hub is not running when you start Skychart, you need to connect from the menu File -> SAMP.

From the same menu you can get the connection status and the list of clients.

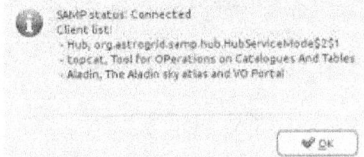

Messages Transmitted

For all this function you have the choice to broadcast the message to every client or to select a single destination client from a list.

Coordinates

To send coordinates use the right click menu SAMP send coordinates.

Table data

You can send a VO table data you get in Skychart from the VO catalog setup. Only VOtable format is supported.

Table selection

When you left click on the chart on an object that is part of a shared VO table the selected object is send as a selection. This is limited to single row selection.

Image

To send a FITS image use the chart right click menu Image list, the "Send image" button.

Messages Received

Remember you can configure which message you want to receive and if you want a confirmation message before to take the action.
Also configure how you want to retain the data when the program is closed. This allow to make a complex data selection using Topcat for example and take the data offline for use with Skychart at the telescope.

Coordinates

Center the chart on the received coordinates and search for an object at this location.
If an object is found you can use all the function available in this case: show detailed information, slew the telescope, ...

Table data

Add the table to the VO catalog list and display the objects on the chart. Only VOtable format is supported.

You can change the symbol, the color or the column selection with the Update button below the list.

Table selection

Mark the selected objects in green on the chart.

Image

Display the image and center the chart the same way as with a DSS image download. Only FITS format is supported.

Example

We want to display on the chart the super-giant stars in the Hyades area using the XHIP catalog as the source of data. If you already know how to work with Topcat you can skip to the last three steps.

We use Topcat to get the data. So launch Topcat and Skychart, connect Skychart to the hub.

On Skychart make a chart centered on the Hyades with a FOV of 20°.

On Topcat click the menu VO → Vizier catalog service. Object name: hyades, click Resolve, enter 10 for the radius (degrees). For "Output columns" select "all".
For Catalog selection, click "By keyword", for "Keyword" enter XHIP, click Search catalogues, select "V/137D" on the list. Click OK.

Send the table data to Skychart: Select the main table V_137D_XHIP in Topcat, open the menu Interop → Send table to .. skychart.

Create the Hertzsprung–Russell diagram in Topcat: open the menu Graphics → Plot, select Table=V_137D_XHIP, X Axis=B-V, YAxis=VMag (not Vmag!), click Flip on Y Axis.

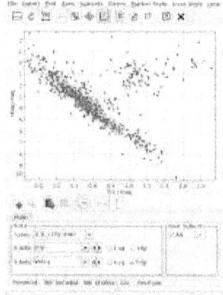

Make a subset with a selection of the red giants branch.
From the graphic menu Subset → Draw subset region, then select the super-giant branch with the mouse.

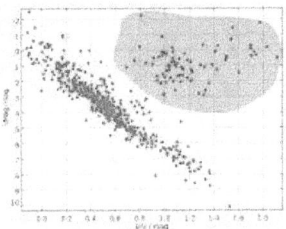

Click the menu Subset → Finish drawing region. Give a name to your subset: "super-giant", select "skychart" and click Transmit Subset.

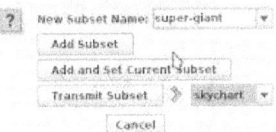

This mark in green all the super-giant on the chart, with the other XHIP stars in red.

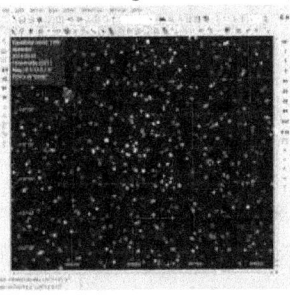

Click on Aldebaran in Skychart, this send a single row selection to Topcat. It highlight the Aldebaran position in the HR plot and in the table browser.

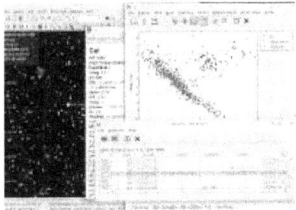

After you get this basic example working you can explore all the powerful function that Topcat can offer to produce the ideal catalog for the data you need. Start by the other options in the VO menu and also try the Joins menu to merge the data from many original tables.

CATGEN

From the menu: **Setup → Catalog**, then in the dialog, click the **CatGen button** on the right.

You can download thousands of Catalog text files with data of any kind of celestial objects from the **CDS [http://cdsweb.u-strasbg.fr/cats/Cats.htx]** or **ADC [http://adc.gsfc.nasa.gov/]** web site, and other sources. Each one contains the information about a few tens to a many millions of objects.

The objective of this program tool is to convert an ASCII catalog text file to a file that SkyCharts can use efficiently. Due to the large amount of available data, it will be impossible to make a choice that will fit all your needs. When you have a specific interest, you can use CatGen to build your own data set from the original professional catalog data. You also can use any new catalog without delay, as soon as it is published.

For a good performance of your newly built catalog in SkyCharts, you can use CatGen to convert the original text based catalog to a catalog in an organized binary format.

Generally, a catalog is a simple formatted text file. In such a catalog, every single line (a record) contains all the data for one individual object. The data of each line consists of separate pieces (fields) of information. These contain the identifier of the object in the catalog, and also determine the properties of this object. The content of every separate piece of data for every line is simply defined by a column position and a length. For example, consider this:

```
1           BD+44 4550      3 36042        46           000001.1+444022000509.9+451345114.44
2           BD-01 4525      6128569                     235956.2-010330000503.8-003011 98.33
3 33    PscBD-06 6357   281285721002I      Var?         000013.0-061601000520.1-054227 93.75
4 86    PegBD+12 5063   87 917012004                    000033.8+125023000542.0+132346106.19
```

At certain fixed positions, there is a fixed type of information. Usually a companion file (Readme) describes the file format with all the necessary details. Detailed information about this file is **here [http://vizier.u-strasbg.fr/doc/catstd.htx]** .

```
Byte-by-byte Description of file: catalog.dat
--------------------------------------------------------------------------------
  Bytes Format  Units     Label      Explanations
--------------------------------------------------------------------------------
   1-  4  I4      ---       HR         [1/9110]+ Harvard Revised Number = Bright Star Number
   5- 14  A10     ---       Name       Name, generally Bayer and/or Flamsteed name
  15- 25  A11     ---       DM         Durchmusterung Identification (zone in bytes 17-19)
  26- 31  I6      ---       HD         [1/225300]? Henry Draper Catalog Number
```

It is important to be familiar with the format of the text based catalog before you try to convert it with CatGen. There are lots of details to pay attention at: units, value ranges (between square brackets), number formats, identifiers, coordinate formats and epochs. A question mark in the Explanations indicate that this data is not always present, look at the data file itself to have an idea of it's frequency. Also be sure to read the notes, when available. A file editor that can handle large files (like **Notepad++ [http://notepad-plus-plus.org/]**) is very useful at this step.

If the data only is available in another format, it is generally easy to write a simple program to extract and format them. A scripting language like Perl can be very useful for that. To handle the CSV files you can use the very good **CSVed [http://home.hccnet.nl/s.j.francke/csved/csvedsetup.exe]**, or **Calc** included with **OpenOffice [http://www.openoffice.org/]**.

You have to select which data to include in your catalog version. For each catalog type, there is a minimal quantity of information that is required to draw the object on the chart. But you can choose to add a number of the other fields with information to display, when you click on such an object.

Remember, the more information you add to the catalog, the more file space will be used on your computer hard disk. It's not allways a good idea to add all the available information, take some time to know what you really need. You can also choose to build two catalog versions, one complete for the desktop computer, and one compact with minimal information for the laptop with limited disk space.

CATGEN Structure

CATGEN has four steps, each on one page:

- **Page 1** Select the input file, name your catalog, and set the catalog type
- **Page 2** Set the general parameters according to the catalog type
- **Page 3** Define the fields to read from the input file to get them included in your catalog
- **Page 4** Set the output options for your catalog and build it

At each step, you can save your project in a **".prj"** file. This contains all the settings you made in CatGen that determine how you want to convert the input file to your catalog. So, you can interrupt building the catalog and continue it later on after loading your .prj file.
The two buttons at bottom of every page in CatGen manage this facility:

- **Save Project** allows you to create (if it didn't already exist) the "xxx.prj" file (xxx = catalog short name).
- **Load project** restores the ".prj" file of your choice.

Page 1

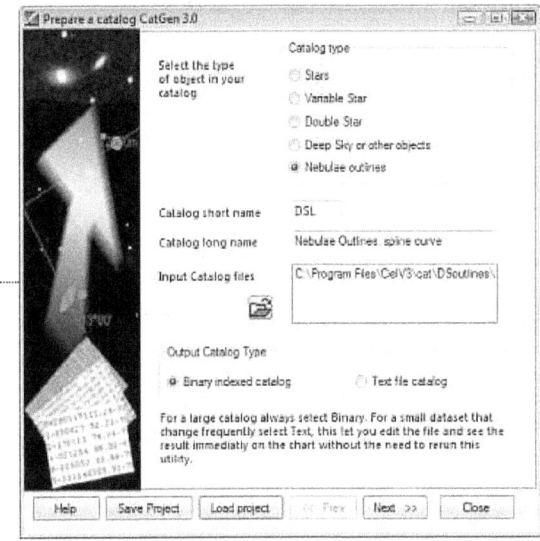

In the first page you need to indicate the **Catalog type** in your catalog, this activates some specific options in the next pages.

The **Catalog short name** is used to identify the catalog. The shortname will be prefixed to the object name in the charts. It must be one to four character long.

The **Catalog long name** is the catalog full identifier.

Click the **Open** button to select the catalog text file. The large catalogs are often split in many smaller files, in such a case select all the files at the same time, the file order doesn't matter.

Choose **Output Catalog Type**. With SkyChart V3, you can choose text file for small catalogs. In this case no data is written out but just a file descriptor that let the program to access your original text file. This way a change to the file is reported to the chart without having to run Catgen again.

If you select "Text file catalog" you also have the option to specify an "Update URL" that point to the latest version of the file on the Internet. This way you can get the last version by a single click in the catalog setup.

Click the **Next** button to go to the second page.

Page 2

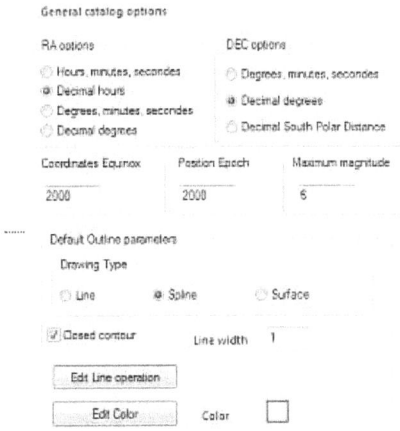

Select the input format of the coordinates:

- **RA options** Hour, minutes, seconds or decimal Hour or Degrees, minutes, seconds or decimal degrees.
- **DEC options** Degrees, minutes, seconds or decimal degrees or South Polar Distance.

Coordinates Equinox is the epoch of the coordinates related to the precession. Use 2000 for ICRS [http://aa.usno.navy.mil/faq/docs/ICRS_doc.php].

Position Epoch is the default epoch position for the proper motion, this date can be changed for each object later in the data file.

Maximum magnitude is the magnitude of the faintest object in this catalog. This is particularly important for the stars catalog to properly set the magnitude scale on the chart.

The second part depends on the catalog type. For stars it is empty. The picture at the right hand side shows **Default Outline parameters** options:

- **Drawing type** (used to connect points in the list) :
  - Line: straight lines
  - Spline: spline curves
  - Surface: fill surface with line color
- **Closed contour** force a closing line between the last point and the first one.
- **Line width** as it says (in pixels)
- **Color** of the line or surface (click on to change it)
- **Edit line operation** (see below) set character strings that will be recognized for drawing operations (comma separated values). Click return when ready.
- **Edit color** (see below) set character strings that will be recognized for lines color (comma separated values). Click return when ready.

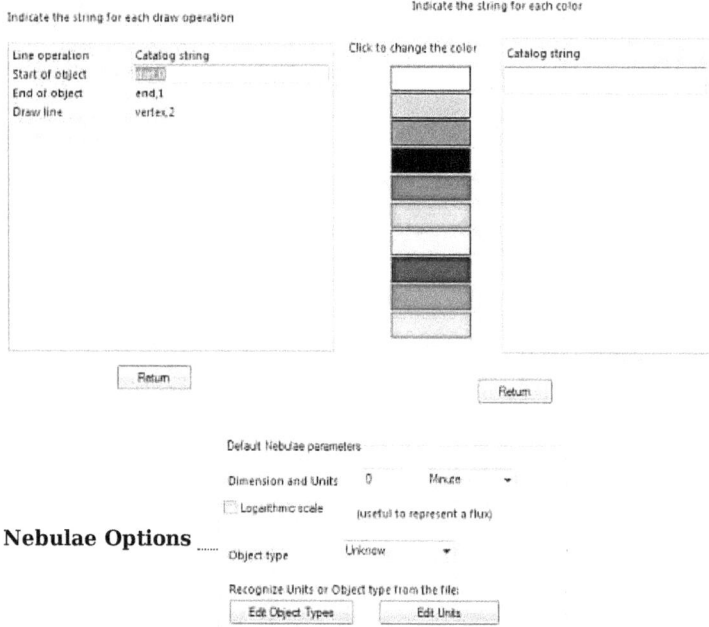

Nebulae Options

- **Dimension and Units** default dimensions (in case of empty fields: no data) and unit.
- **Logaritmic scale** check this box if the field used for size is a flux, i.e. for a radio source catalog.
- **Edit Object Type** set character strings that will be recognized to set the object type (comma separated values). Click return when ready.
- **Edit Units** set character strings that will be recognized to set the unit size (comma separated values). Click return when ready.

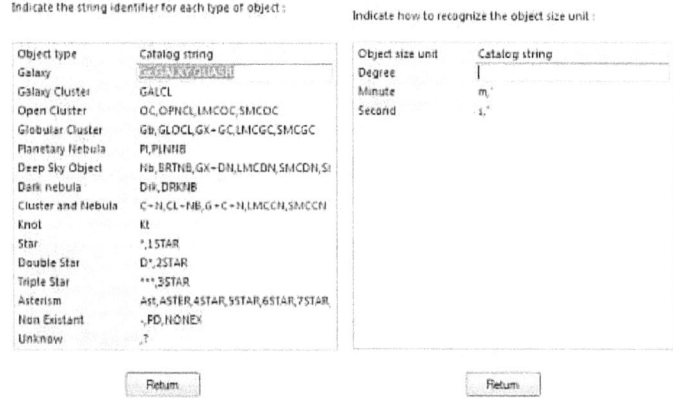

Click the **Next** button to go to the third page.

Page 3

This page enables you to map the data from the input file to your catalog file. The list shows different fields, depending from the catalog type you chose in the first page.

Select each field you want to include in you catalog and define the position in the sample record.

Required fields are enclosed between square brackets "[]", required units are between brackets "()".

Indicate in **Label** the label you want to show for this field. **First Char** and **Length** describe the position of the field for each line in the input file. You can type the value here (obtained from the catalog Readme file) or mark the data in the sample record with the mouse. When you use your mouse, be sure the field length you mark will be long enough to contain the longest string of field information in the catalog file.

Field list for each catalog type:

Required fields:

Stars	Variable Stars	Double Stars	Nebulae	Nebulae Outlines
RA	RA	RA	RA	RA
DEC	DEC	DEC	DEC	DEC
Magnitude (V)		Magn. comp. 1		Line operation
		Separation		

Fields used to draw the charts:

Stars	Variable Stars	Double Stars	Nebulae	Nebulae Outlines
Catalog ID	Catalog ID	Catalog ID	Catalog ID	Catalog ID
B-V	Magnitude Max.	Position angle	Nebula type	Line width
Proper motion RA	Magnitude Min.		Magnitude	Line color
Proper motion DEC	Magnitude code		Surface brightness	Drawing type
Position Epoch			Largest dimension	
Magnitude B			Smallest dimension	
			Dimension unit	
			Position angle	

Fields for information only:

Stars	Variable Stars	Double Stars	Nebulae	Nebulae Outlines
Magnitude R	Period	Magn. comp. 2	Radial velocity	Comment
Spectral class	Variable type	Epoch	Morphological class	String..
Parallax	Maxima Epoch	Component name	Comment	Numeric..
Comment	Rise Time	Spectral class comp. 1	String..	
String..	Spectral class	Spectral class comp. 2	Numeric..	
Numeric..	Comment	Comment		
	String..	String..		
	Numeric..	Numeric..		

At the bottom of the list you find ten strings and numeric values you can use freely for any data.

If you check **Use this field name as an Alternate name** the string value can be used as an alternate name for the object. This name is used in the case the main name (Catalog Id) is missing or it can be added to the index file of the catalog.

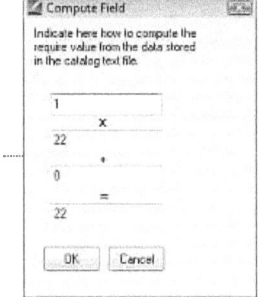

The **Advanced** button enables you to compute a linear transformation of the catalog data. It is active only for numeric values. A use can be to obtain the object coordinates when these aren't directly available from the catalog.
If the catalog gives the object position in millimeters in the plate from the galaxy center (i.e. ra: 0h42m42.00s dec:+41°16'00.0") and you know the top of the plate is at North and the plate scale is 2 seconds per millimeter.
The approximate (without plate correction) coordinates in degrees are:

```
DEC = 41.26667 + Xmm * (2/3600)
RA = 10.67500 - Ymm * (2/3600/cos(DEC))
```

Select "Decimal degree" for both RA and DEC, indicate the position in millimeters and set the "Advanced" value to:

```
for RA : -0.00073837 and 10.67500
for DEC : 0.00055555 and 41.26667
```

If the data requires some more complexe transformations you must do the transformations before running Catgen. A Perl script can solve almost any case.

Click the Next button to go to the fourth page.

Page 4

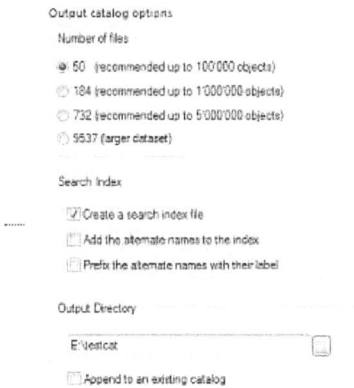

In the last page you set the options for the generated catalog.

Set the number of files as recommended depending the number of objects in the catalog, this is very important to obtain good performance.
Check the **create a search index file** box if you want to create a search index to search objects in this catalog by their names.
Check the **Add the alternate names to the index** box if you want the fields you defined as alternate names to be included in the index.
Check the **Prefix the alternate names with their label** box if you want to prefix the alternate names with the label you set for the corresponding field. I.e. if a column of the input file list the UGC catalog number, you can set the index value to UGC7442 instead of 7442, which can be confusing.
Choose an **Output directory** to save the catalog files. It is a good habit to use a separate directory for each catalog. Ususally these directories are subdirectories of [SkyCharts installation directory]/cat/
Check the **Append to an existing catalog** box if you want to append the data to an existing catalog of the same name in the same directory.
This is useful to create a single catalog from various sources. In this case the catalog structure must be exactly the same as the existing catalog, otherwise the data is lost, please make a backup before to try this option.

Now you're this far, save your job definition with the **Save Project** button, to enable the possibility of easy changes. Now you can press the **Build Catalog** button!

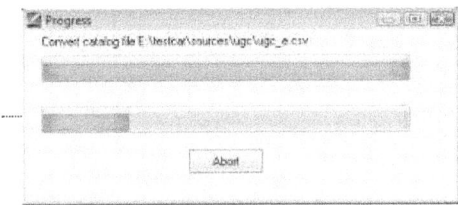

A progress box shows you the current operation. You can interrupt the process by pressing the **Abort** button. In such a case the catalog will not be usable.

After the build is complete, the progress box is closed. You can exit the program with the **Close** button.

In case a catalog file record doesn't contain a valid numeric value for a field, this object will not be included in the catalog. Those cases find their way to a file called **reject.txt**.

It's a good idea first to test your definitions with a limited number of records to avoid a file consisting of millions of rejected lines! You can find a Windows equivalent of the Unix commands "head", "tail" and "grep" in a package **here [http://sourceforge.net/project/showfiles.php?group_id=9328&package_id=9393]** that will help you to extract a few lines from a large file.

If all worked well, you are now ready to use the catalog. Refer to **SkyChart documentation [http://www.ap-i.net/skychart/en/documentation/catalog#catalog]** to activate the catalog.

Notes for Linux and Mac users

Please read carefully if you plan to make a big catalog with 9537 files.

As it read lines from the input files Catgen need to randomly write the data to the 9537 output files, depending of the star coordinates. So the 9537 files need to be open at the same time.

As Linux and Mac OS are multi-user operating system there is safe-guard to avoid one user to eat all of the system resource. For a standard Linux installation for example the maximum number of file a user can open at the same time is limited to 1024.

On Mac OS X the way to do may vary depending on your version. See the first answer to this question [http://unix.stackexchange.com/questions/108174/how-to-persist-ulimit-settings-in-osx-mavericks].

On Linux you first need to change the system limit in the file limits.conf :

```
sudo vi /etc/security/limits.conf
```

add this two lines at the end of the file:

```
* soft nofile 1024
* hard nofile 32768
```

I recommend you keep the default value of 1024, this can be a good protection again a runaway process.

Save the file, and logout from the system as this file only apply at next login.

Then open a terminal window and type :

```
ulimit -S -n 20000
```

and from the same terminal type :

```
skychart
```

This is very important as the limit change is only for the current instance of the shell, not system width.

Tool box editor

This window is show when you click the Script button at the bottom of a Tool box. It allow to configure and program the action for the tool box.

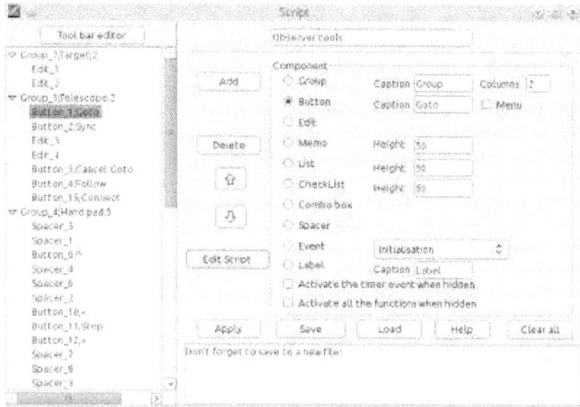

Create the tool layout

First set the title for your tool in the top text field.

Then you need to add at least one Group: click the Group button, set the group title and the number of column element you want on each row. Then click the Add button.

Then use the same principle to add the elements you want, for example a text Edit and a Button.
This elements can only be added to a group.
A name is automatically assigned to each element, for example Edit_1 . This is this name you need to use to access the element properties from your script.
For some elements you can select a title or a height in pixel.
To change the number of elements per row, create a new group with the required value but without title.

Special elements are the Events. This not add anything visible on the screen but allow you to write the require script to process an event.
Event are :

- Initialization: Run once after you click the Apply button, or when the program start.
- Activation: Run every time this tool box is show.
- Timer: Run at periodic interval, enter the interval in second when you select this event.
- Telescope move: Each time the telescope position change.
- Chart refresh: Each time the chart redraw.
- Object identification click: Each time and object is select by a click on the chart or as the result of a search.
- Distance measurement: When you measure a distance on the chart.
- Telescope connected: When the telescope is connected.
- Telescope disconnected: When the telescope is disconnected.
- Translation: Run after initialisation or when you change the program language.

The events are normally inactive when the tool is not show on the screen. One exception is the Timer event you can always activate by checking the corresponding box. But be careful with the performance issue this option can imply.
Use the Activation event to update your tool with the change that can occur on the map when the tool was inactive.

You can reorder the group or elements by using the vertical arrows or by drag and drop in the left tree.

If you select an element in the tree and change one of it's property the Update button appear to allow to apply the change.

A script can be attached to following elements: Button, Menu, Combo box, Event.
When you select one of this elements in the tree the "Edit script" button appear. See below for the details.
You can add a function to the right click menu of the chart. To do that create the function for a button and check "Menu".

When you are ready click the Apply button to show your elements on the tool box and compile the scripts.
Don't forget to save your work to a file with the Save button (default file extension is .cdcps).
Also save the program configuration to have you tool box automatically loaded the next time you start the program.

Script editor

The editor allow to write the script itself that execute on a button click, a combobox selection, or an event.

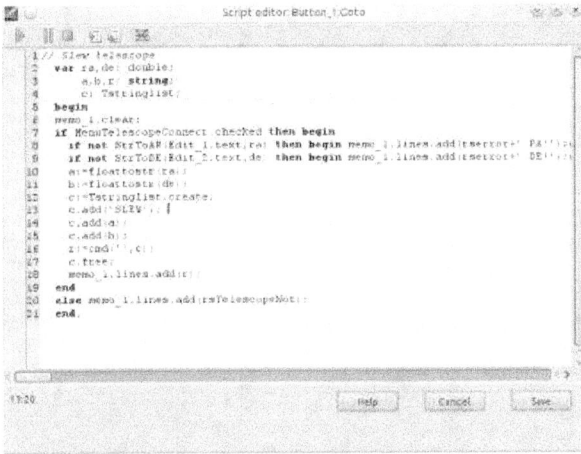

The language to use is Pascal Script [http://en.wikipedia.org/wiki/Pascal_Script], based on Object Pascal [http://en.wikipedia.org/wiki/Object_Pascal].

Define first the global variables, then the procedure and function if any, then the private variable, and finally the main code start with **begin** and end with **end.**

Read the script example page for a quick start.

As the each script is specific to one event there is no possibility to have global variables across scripts. For example you cannot set a variable when a button is pressed and later use this variable in another button click.

To solve this issue a number of global variables are predefined in the main program and specific functions allow to write and read them.

This functions and other specific to the interface with Skychart, they are described in a separate script reference page.

Use the **Save** button to record your change and return to the main window.

The top button are related to the debugging function as describe below.

Script debugger

Simple debugging function are available to test your code.

To run the script in debug mode press the green arrow **Run** button. The program is first compiled.

In case of compilation error, the corresponding row is highlighted in yellow, and the error message is show in the bottom message area.

If the compilation finish without error the program start to run and stop on the first code line of the main procedure. The current execution position is highlighted in blue.

You can now use the **Step over** button to execute your program line by line.

The **Step into** button do the same, except if the current line is a call to one of your function. In this case **Step into** allow to run the function line by line, but **Step over** execute the function and stop at the main program next line.

You can also set a breakpoint on a specific line to jump directly at this position.

To set a breakpoint click on the leftmost column to show a red icon.

Use the Run button to jump to the next breakpoint. The current line is then highlighted in red.

You can remove a breakpoint by clicking on the red icon or all at at time with the **Remove all breakpoint** button.

You can display the value of variables when the program is in pause at a breakpoint or after a **Step over** click.
Just click on the variable name anywhere in the program source to display the value in the message area.
Note this work only for local variables, not for object properties.

You can use the **Pause** button to pause the program execution. This can be useful to examine the condition of an infinite loop for example.

The **Stop** button terminate the program execution immediately.

Add elements manually

From the menu: **Settings→ Solar Sytem... → Comet → Add**
or
From the menu: **Settings→ Solar Sytem... → Asteroid → Add**

It can be helpful to add manually elements to your existing list of comets or asteroids. You also can display one asteroid amidst the stars by deleting the position data of the other asteroids and entering the element data for your single asteroid.

Warning

Elements are used to define conical section orbits like ellipses of asteroids and ellipses and parabolas of comets. The resulting calculated positions are only useful for objects that are not close to planets. In the case of Near Aarth Objects (NEOs) that are close to the Earth, the calculated postions based on the elements are totally useless. If you want to display the postion or path of a NEO with Cartes du Ciel, you can do this based on a list of calculated postions and adding this list to your catalogs. **Here** you find instructions on how to do that.

Requesting elements

You can request elements from **NASA JPL Horizons**. Open this link in a new window: http://ssd.jpl.nasa.gov/horizons.cgi [http://ssd.jpl.nasa.gov/horizons.cgi]
Copy the settings below, of course with your adjustments. Select for *Ephemeris Type* `Orbital Elements`. Choose for *Target Body* your demanded object. You need to set *Center* as `@sun`. Set at *Time Span* the time range for which you want to retrieve the elements.

Ephemeris Type	[change]	: ELEMENTS
Target Body	[change]	: Asteroid 5610 Balster (2041 T-3)
Center	[change]	: Sun (body center) [500@10]
Time Span	[change]	: Start=2014-08-09, Stop=2014-08-10, Step=1 d
Table Settings	[change]	: defaults
Display/Output	[change]	: default (formatted HTML)

Entering the data

Data from the Object Data page

Enter a clear designation of the object. In this example I choose "Asteroid 5610 Balster (2041 T-3)".
In the part *Physical Parameters* you find the value **H= ...**. Copy this as *H absolute magnitude*.
Copy the value **G= ..** as *G Slope parameter*.

The Results data

Further, in the **Results** part we find more data to enter. The results list shows data in mathematical notation. Cartes du Ciel can process this data both in this form as in decimal notation.

```
$$SOE
2456878.500000000 = A.D. 2014-Aug-09 00:00:00.0000 (CT)
EC= 4.410631654154605E-02 QR= 2.680223611595490E+00  IN= 3.348991791769490E+00
OM= 2.437396671671045E+02  W = 3.329253281089416E+02 Tp= 2456381.044945736416
N = 2.099239155691789E-01 MA= 1.044277128106851E+02 TA= 1.092504742118517E+02
A = 2.803893003977550E+00 AD= 2.927562396359610E+00 PR= 1.714907036789549E+03
2456879.500000000 = A.D. 2014-Aug-10 00:00:00.0000 (CT)
EC= 4.410698013669621E-02 QR= 2.680219915877766E+00  IN= 3.348991402148423E+00
OM= 2.437396179792080E+02  W = 3.329246715210186E+02 Tp= 2456381.042354188394
N = 2.099241311623329E-01 MA= 1.046382882182052E+02 TA= 1.094556083560238E+02
A = 2.803891084235605E+00 AD= 2.927562252593444E+00 PR= 1.714905275571270E+03
$$EOE
```

Take note of the abbreviations and copy the corresponding data:

1. 2456338.500000000 as value for **Epoche (JD)** (on top, directly under $$EOE)
2. **MA**: 1.044277128106851E+02 as value for **Mean anomaly**
3. **W**: 3.329253281089416E+02 as value for **Argument of perihelion**
4. **OM**: 2.437396671671045E+02 as value for **Longitude of ascending node**
5. **IN**: 3.348991791769490E+00 as value for **Inclination**
6. **EC**: 4.410631654154605E-02 as value for **Excentricity**

7. **A**: 2.803893003977550E+00 as value for **Semimajor axis**

At *Reference* enter the reference to the source of which you retrieved your data. In this example **"Horizons"**.
At *name* you enter the name that will show up in the label. To discriminate between previously calculated positon data it may be useful to enter your object name extended with some sort of extra information.

Click at the button **Add**. If you entered all data in a format that CdC can process, you will recieve an **"OK!"**.

Now, do a new *prepare monthly data*, compute the postitions according these instructions.

Display of NEOs

It is possible to display Near Earth Objects (NEOs) in Skychart after meticulous calculation of their positions on the sky. To do these calculations, we need our position on Earth and the time for moment of display. We can obtain the results of these calculations from the Minor Planet Center [http://www.minorplanetcenter.net/iau/MPEph/MPEph.html] or from Horizons website from NASA JPL [http://ssd.jpl.nasa.gov/horizons.cgi]. After that, we can copy the data in a ASCII text file and use CatGen to compile a catalogue.

We can not use the orbital elements to do a trustworthy track display of a NEO flyby!

It is strongly advised to do a latest calculation and compilation of the catalog short before observing. The track of the NEO is likely to be different from the earlier predictions!

Why we're not using orbital elements now

Usually Skychart is very well capable to create reliable charts with comets and asteroids based on orbital elements. Because of the usual large distances, the perturbations caused by the gravity field of the Earth or other planets are only very small influences on the track of the solar system objects. In case of a flyby situation, the gravity of the Earth is one significant perturbator extra, it has a large influence on the trajectory.

Orbital elements are mathematical data to determine conical section orbits (ellipses, parabola's) of solar system objects moving around the Sun. The orbital elements are calculated based on two bodies, the object itself and the Sun. Of course, there are the perturbations by the planets, but distances are very large, so influences are little. This is why we -under normal conditions- can use orbital elements to make reliable predictions for some time.

In case of NEOs, these normal conditions are not the case. To make predicitons about time and position, we need to do calculations based on three bodies: The object itself, the Sun and the Earth. Under the extra influence of Earths gravity, the track of the object does not follow its former conical section orbit anymore. Because orbital elements describe only conical sections, these aren't helpful in case of NEOs. If you would hold on to use orbital elements during a flyby event, you wil notice great changes of the orbital element data over short periods of time. Only after the object has moved far enough from Earth's influence you can rely on the orbital elements again.

Download of the CatGen project-files

You can use ephemerides calculations from both the Minor Planet Center (MPC) and the NASA JPL Horizons, or you can choose one of them. Skychart is capable to display catalogues of both sources in one single chart.

For our convenience, I compiled a zip file containing:

MPC.prj	The CatGen project file for the efemerides-format of MPC
Horizons.prj	The CatGen project file for the efemerides-format of NASA JPL Horizons
mpc.txt	Raw ASCII-text file as an example, containing efemerides from MPC
horizons.txt	Raw ASCII-text file as an example, containing efemerides from Horizons
mpc.hdr	Example-catalogue header, ready for usage at 2013-02-15, generated with CatGen, based at MPC data.
horizons.hdr	Example-catalogue header, ready for usage at 2013-02-15, generated with CatGen, based at Horizons data.
mpc.info2	Example-catalogue data, ready for usage at 2013-02-15, generated with CatGen, based at MPC data.
horizons.info2	Example-catalogue data, ready for usage at 2013-02-15, generated with CatGen, based at Horizons data.
README.txt	Short description of the purpose of these files.

First of all, download this zip file [http://www.centaurus-a.nl/home/files/ephemerides.zip] and save it in the directory where you want to store your catalogues. With windows, the standard catalogue directory is `C:\Program Files\Ciel\cat`, with Linux that is `/usr/share/apps/ciel`. Unzip the file in your catalogue directory. After that, you will find a new directory named `ephemerides`. It will contain the above-mentioned files.

Retrieval of Minor Planet Center data

Open the source of the MPC data by this link [http://www.minorplanetcenter.net/iau/MPEph/MPEph.html] in a new Window. Select the radio button `Return Ephemerides`, in case it wasn't selected. Enter the object identification(s) in the rectangular window for which you want generate the catalogue, for example **2012 DA14**.
Select of enter at the same page also:

- The wished start date for the calculations, in Universal Time. Several formats are allowed, for example
  - `2013 02 15 194600` is 07:46:00 PM UT (h:m:s) at february 15th 2013
  - `2013/02/15.75` is february 15th 2013 at 6:00 PM UT.
- The number of positions that you demand to be calculated (`number of dates to output`).
- Optional: enter a number in the text box (`Ephemeris interval`) if you want to an interval for the number of time units. You will select this time unit next at *Ephemeris units*.
- Your wished time unit (`Ephemeris units`) in days, hours, minutes or seconds.
- In decimal notation: (`longitude`), the (`latitude`) and the (`altitude`) above sea level in meters of the site of your observations.
- Keep selected: `full sexagesimal, total motion and direction, "/min.`
- Select according to your preference `Measure azimuths`.
- Keep selected: `Format for elements output: none.`
- Click at the `Get ephemerides/HTML page` button.

```
2013 02 15 180000 11 46 49.9 -59 00 03   0.00036 0.988 105.1 74.9   9.6 1164.89   006.0   128 -54  -11  0.30  128 +45   54 078.6 / Map / Offsets
2013 02 15 180100 11 47 05.7 -58 40 38   0.00036 0.988 105.4 74.6   9.8 1178.23   006.0   127 -54  -11  0.30  129 +45   54 078.5 / Map / Offsets
2013 02 15 180200 11 47 21.4 -58 21 00   0.00036 0.988 105.7 74.3   9.7 1191.74   005.9   127 -54  -12  0.30  129 +45   54 078.4 / Map / Offsets
2013 02 15 180300 11 47 36.9 -58 01 07   0.00036 0.988 105.9 74.0   9.7 1205.47   005.9   126 -53  -12  0.30  129 +45   54 078.2 / Map / Offsets
2013 02 15 180400 11 47 52.4 -57 41 01   0.00035 0.988 106.2 73.8   9.7 1219.36   005.8   126 -53  -12  0.30  129 +45   54 078.1 / Map / Offsets
2013 02 15 180500 11 48 07.7 -57 20 41   0.00035 0.988 106.5 73.5   9.7 1233.43   005.7   125 -53  -12  0.30  130 +44   54 078.0 / Map / Offsets
2013 02 15 180600 11 48 22.9 -57 00 07   0.00035 0.988 106.8 73.2   9.7 1247.67   005.7   125 -53  -12  0.30  130 +44   54 077.9 / Map / Offsets
2013 02 15 180700 11 48 38.0 -56 39 18   0.00035 0.988 107.1 72.9   9.6 1262.14   005.6   124 -52  -12  0.30  130 +44   55 077.8 / Map / Offsets
```

Now open in the earlier created `ephemerides` directory the file MPC.txt using an ASCII-editor (i.e. notepad of vi). Select in the browser containing the generated MPC-efemerides only those lines that contain position data. Copy the lines in the MPC.txt document and save it. We are going to use this document as an ***Input catalog file*** with CatGen.

If you don't want to use ephemerides data from Horizons you can savely skip the next paragraph and continue with Genereration of an ephemerides catalogue with CatGen.

Retrieval of Horizons position data

Open this link [http://ssd.jpl.nasa.gov/horizons.cgi] in a new window.
Adapt your settings like the example below.

- Select for **Ephemeris Type** `Observer Tables`.
- Enter your demanded object at **Target Body**.
- Click under **Specify Observer Location:** at the line `specify latitude, longitude, and altitude` and set the coordinates and altitude above sea level at **Specify Observer Coordinates**. After that, click the button **Use specified Coordinates**.
- Next, set the time range for which you want to retrieve the position data.

Ephemeris Type	[change]	: OBSERVER
Target Body	[change]	: Asteroid (2012 DA14)
Observer Location	[change]	: user defined (5°52'07.0"E, 51°49'24.0" N, 60 m)
Time Span	[change]	: Start=2013-02-15 19:00, Stop=2013-02-16 02:00, Step=10 m
Table Settings	[change]	: defaults
Display/Output	[change]	: default (formatted HTML)

All demanded data are entered. So now, click the button **Generate Ephemeris**. In the **Results** you will find the positions like the example below.

```
2013-Feb-15 19:00  m   12 00 04.31 -31 04 57.4     8.32 0.00025777599773   -3.0499410 127.3788 /L   52.6083
2013-Feb-15 19:01  m   12 00 15.84 -30 26 57.6     8.30 0.00025656475249   -2.9902275 127.8449 /L   52.1423
2013-Feb-15 19:02  m   12 00 27.35 -29 48 37.0     8.27 0.00025537764519   -2.9295700 128.3127 /L   51.6746
2013-Feb-15 19:03  m   12 00 38.84 -29 09 55.8     8.25 0.00025421505430   -2.8679690 128.7818 /L   51.2056
2013-Feb-15 19:04  m   12 00 50.31 -28 30 54.3     8.23 0.00025307735783   -2.8054258 129.2522 /L   50.7354
```

Now open in the earlier created `ephemerides` directory the file `horizons.txt` using an ASCII-editor (i.e. notepad of vi). Select in the browser containing the generated MPC-efemerides only those lines that contain position data. Copy the lines in the horizons.txt document and save it. We are going to use this document as an ***Input catalog file*** with CatGen.

Generation of an ephemerides catalogue with CatGen

If you keep the default file name for the data and you not want to do any change to the provided example mpc.hdr or hori.hdr you can skip this section and go directly to Activate the catalogue.

For common documentation on the usage of CatGen click here.

From the menu: **Setup → Catalog**, then the tab **Catalog** in the dialog, click the **CatGen button** at the right hand side.

CatGen Page 1

Click at the bottom of the new dialog the button **Load project**. Find the directory were you earlier unzipped the downloaded file and select the appropriate **.prj file** that belongs to the ephemerides source.

Replace at **Catalog short name** the identification for a short, recognizable catalogue name, max. 4 signs. For example, the last four signs from the object name.

Enter at **Catalog full name** the object identification combined with the date.

Click the button **Input catalog files** to set the input-catalog. Select the .txt file with the ephemerides data that belongs to the project file you using now.

Leave the output catalog type untouched to **Text file catalog**.

Click **Next**.

CatGen Page 2

For MPC: Leave equinox to 2000, set epoch to the year of the date for which you're creating the chart.

For Horizons: Find in the results part of the Horizons page the equinox and epoch data. These are displayed above the coordinates table. It is likely that you will find data as `Initial FK5/J2000.0 heliocentric ecliptic osculating elements`, in that case, leave equinox to 2000. The epoch has been given in Julian format and as ordinary date. Take for the epoch the year part of the ordinary date.

Search in the results table which maximal magnitude we can expect from the object. Round the value up, and enter it in the magnitude field.

At **Object type** you can determine the displayed shape. Choose a simple, well recognizable shape.

Click **Next**.

CatGen Page 3

Here, all needed settings are in place for the formats of MPC - or Horizons data. Click **Next**.

CatGen Page 4

Now, set the path where you want to write the catalogue file. A logical place is the same directory where you placed the ASCII and .prj files.
Save your project settings by a click at the button **Save project**.
Now click the button **Create catalog**, after that, click **Close**.

Activate the catalogue

From the menu: **Setup → Catalog**, then the tab **Catalog** in the dialog, click at the button **Add** in order the tell Skychart where to find your new catalog. Search for the `.hdr` file of the catalogue in the directory. After that, activate the catalogue by a click at the red dot. To confirm activation, it will turn green.

Display

If you didn't change the other settings, the position of the asteroid will be displayed by a red lozenge. The catalogue will respond as if it is a deep sky catalogue, so you can set the magnitude of the stars independently from the asteroid.

In order to display the NEO in the same coordinate system that we retrieved from Horizons, check your settings: **Setup → Chart, coordinates...**, choose under **Type of coordinates** the radio button `Astrometric J2000 (mean equinox J2000, epoch of the date)`.

Keyboard Shortcuts

Scale, Field of Vision

- **+**: Zoom in
- **-**: Zoom out
- **1, 2, ...9, 0, a**: predefined FOV range #
- **Mousewheel push**: Zoom in
- **Mousewheel pull**: Zoom out
- **Continuous left mouse click with drag of the mouse cursor**: select a zone to Zoom in
  - **Continuous left mouse click with drag of the mouse cursor**: move the selected zone over the chart
  - **left-click in the selected zone**: confirm the Zoom

For a precise form of control over the Field of Vision, see the lower part of **View → Postion**.

The left mouse click behavior can be changed by using the Change mouse mode button

Directions and displacements

- **n**: Display the North horizon
- **e**: Display the East horizon
- **s**: Display the South horizon
- **w**: Display the West horizon
- **z**: Display the Zenith
- **keyboard and numpad arrows (even diagonal)**: chart displacement in the direction of the arrow
  - **+ Ctrl**: faster displacement
  - **+ Shift**: slower displacement
- **Mousewheel click and drag**: chart displacement
- **shift + left mouse key pressed and drag**: chart displacement

For a precise positioning of your chart, see the upper part of **View → Position**.

Finder mark

Enter the rotation mode

- **Shift+C** : Rotate the main camera rectangle.
- **Shift+G** : Rotate the guider position around the main camera center.
- **Shift+S** : Rotate both the main camera and guider at the same time.

Rotate

- **Left arrow** : Rotate counterclockwise by five degrees.
- **Right arrow** : Rotate clockwise by five degrees.

Quit the rotation mode using the same key as to enter

Telescope

- **Ctrl+K** : Abort slew

Stars

The following is only applicable when the display of stars is set to **parametric mode**.

- **ctrl + q**: increase faint stars size
- **ctrl + a**: decrease faint stars size
- **ctrl + w**: increase brightness
- **ctrl + s**: decrease brightness
- **ctrl + e**: increase contrast
- **ctrl + d**: decrease contrast
- **ctrl + r**: increase color saturation
- **ctrl + f**: decrease color saturation
- **ctrl + i**: show/hide the background pictures

Object information and label

- **left-click on an object**: select the object, show label
- **right-click on an object**: object pop-up menu
- **Ctrl+L**: Display the chart informations.
- **Ctrl+Shift+L**: Display the chart legend.

While "**edit label**" mode is on:

- **left-click on label**: detailed information
- **right-click on label**: edit label pop-up menu

Window

- **F1**: contextual help (These documents)
- **F11**: full screen
- **ctrl + b**: show/hide toolbars
- **ctrl + c**: copy chart to clipboard
- **ctrl + l**: reload the translation file for the current language
- **ctrl + tab**: activate the next chart

Command line options

Cartes du Ciel - SkyChart accepts the following options on the command line:

Option	Parameter	Function
--config	configuration file path	Lets you specify the configuration file to use instead of the default "%LOCALAPPDATA%\skychart\skychart.ini" or "~/.skychart/skychart.ini"
--loaddef	option_file_name	Same as the LOADDEFAULT server command Use this option to load an extract of the configuration file that temporarily replace some option from skychart.ini.
--unique		Do not launch the program if another instance is already running but instead send the other options to the running instance
--quit		Use this option in conjunction with --unique to close the running instance, or to exit immediately.
--nosplash		Do not show the splash screen on startup
--daemon		Start the program in background without showing the main window
--nosave		Do not save options on exit. Useful with --loaddef option
Options affecting the first active chart		
--load	saved_file_name	Same as the LOAD server command Use this option to load a template with all the chart setting you cannot set here.
--search	object_name	Same as the SEARCH server command
--setproj	ALTAZ/EQUAT/GALACTIC/ECLIPTIC	Same as the SETPROJ server command
--setfov	00d00m00s or 00.00	Same as the SETFOV server command
--setra	RA:00h00m00s or RA:00.00	Same as the SETRA server command
--setdec	DEC:+00d00m00s or DEC:00.00	Same as the SETDEC server command
--setobs	LAT:+00d00m00sLON:+000d00m00sALT:000mOBS:name	Same as the SETOBS server command
--settz	Etc/GMT	Same as the SETTZ server command
--setdate	yyyy-mm-ddThh:mm:ss or "yyyy-mm-dd hh:mm:ss"	Same as the SETDATE server command
--setcat	"path shortname active min max "	Same as the SETCAT server command
--resize	"width height "	Same as the RESIZE server command
--dss		Same as the PDSS server command
--saveimg	"PNG/JPEG/BMP filename quality"	Same as the SAVEIMG server command
--print	"PRT/PS/BMP PORTRAIT/LANDSCAPE COLOR/BW filepath"	Same as the PRINT server command

Warning, contrary to commands sent by TCP/IP, options and parameters are separated by the "=" sign without any blank space. If the parameter contains spaces it must be enclosed in double quotes "".

On Unix you can send a signal to the running program with the kill command:

Signal	Action
1, HUP	Reload the default chart and options
15, TERM	Gracefully close the program

Automation example using command line

Automation is normally done by sending server commands to the TCP/IP connexion but in some case it may be simpler to use the command line only.

First star the main instance :
skychart --unique

Set the image size you want:
skychart --unique --resize="1024 768"

Then the following command show a chart of Messier 1 without opening a new window:
skychart --unique --setfov=3 --setproj=EQUAT --search=M1

Save an image with the current chart:
#skychart --unique --saveimg="PNG /tmp/m1.png"

You can now repeat with other objects or FOV. When finished you can close the main window:
skychart --unique --quit

It is also possible to do all at a time and exit, without displaying anything on the screen, this is useful if you want only one image:
#skychart --nosplash --daemon --resize="1024 768" --setfov=3 --setproj=EQUAT --search=M1 --saveimg="PNG /tmp/m1.png" --quit

Server Commands

SkyChart can work as a server. It accepts the following commands from a TCP/IP connection:

Connection

You can use any socket library or object to connect to Skychart from your software, or use the commands telnet or netcat from a script.
For examples with various languages, see http://skychart.svn.sourceforge.net/viewvc/skychart/trunk/skychart/sample_client/ [http://skychart.svn.sourceforge.net/viewvc/skychart/trunk/skychart/sample_client/]

The standard port is 3292, this can be changed by the user in the configuration menu. The program can also choose to listen to another random port if the configured port is busy.

Skychart maintains a file with information about the port and it's status.
In Linux and Mac: [User settings]/tmp/tcpport
In Windows: Registry key HKCU\Software\Astro_PC\Ciel\Status\TcpPort

The file (or key) do not exist if Skychart was never run.
It contains 0 if the program is not running, or the server is disabled in the configuration menu.
It contains the listen port if the program is ready to accept a connection.

Main Commands

Command	Parameters	Comment or GUI Equivalent
NEWCHART	chart_name	File → New Chart
CLOSECHART	chart_name	File → Close Chart
SELECTCHART	chart_name	Window → chart_name
LISTCHART		Window
SEARCH	object_name	Search tool (Main Bar)
GETMSGBOX		Returns the status bar content
GETCOORBOX		Returns the coordinates shown in the status bar
GETINFOBOX		Setup → Display → Labels - Display the chart information in the menu bar
FIND	object_class object_name	Same as search, but lets you specify the class of object you want: 0=nebula, 1=na, 2=star, 3=star, 4=variable, 5=double, 6=comet, 7=asteroid, 8=planet, 9=constellation, 10=line catalog
SAVE	saved_file_name	Save the current chart to the specified file
LOAD	saved_file_name	Load the chart from the file
LOADDEFAULT	option_file_name	Use this command to load an extract of the configuration file that temporarily replace some option from skychart.ini.
SETCAT	path shortname active min max	Add or change a Catgen catalog. The fields are the same as in the catalog setup.
SHUTDOWN		Close the program
RESET		Reload the default chart and options (same as signal HUP)
?		No GUI equivalent, list of available commands

Chart Commands

Command	Parameters	Comment or GUI Equivalent
ZOOM+		View → Zoom in
ZOOM-		View → Zoom out
MOVEEAST		Keyboard Left arrow
MOVEWEST		Keyboard Right arrow
MOVENORTH		Keyboard Up arrow
MOVESOUTH		Keyboard Down arrow
MOVENORTHEAST		Numpad 7 (Ver num off)
MOVENORTHWEST		Numpad 9 (Ver num off)

MOVESOUTHEAST		Numpad 1 (Ver num off)
MOVESOUTHWEST		Numpad 3 (Ver num off)
FLIPX		Chart > Tranformation > Mirror Horizontally
FLIPY		Chart > Tranformation > Mirror Vertically
SETCURSOR	pixX pixY	
CENTRECURSOR		Right-clic → Center
ZOOM+MOVE		Right-clic → Zoom + Move
ZOOM-MOVE		Right-clic → Zoom - Move
ROT+		Chart → Tranformation → Rotate Right
ROT-		Chart → Tranformation → Rotate Left
SETEQGRID	ON/OFF	Chart → Lines, Grid → Add Equatorial Grid
SETGRID	ON/OFF	Chart → Lines, Grid → Show Coordinate Grid
SETSTARMODE	0/1/2	Setup → Display → Display → Star Display
SETNEBMODE	0/1	Setup > Display → Display → Deep Sky Objects Display
SETAUTOSKY	ON/OFF	Setup → Display → Sky Colour
UNDO		Edit → Undo
REDO		Edit → Redo
SETPROJ	ALTAZ/EQUAT/GALACTIC/ECLIPTIC	Chart → Chart Coordinate System
SETFOV	00d00m00s or 00.00	Icon (FOV input fields) or icon (Main Bar)
SETRA	RA:00h00m00s or RA:00.00	Icon (RA input fields)
SETDEC	DEC:+00d00m00s or DEC:00.00	Icon (DEC input fields)
SETOBS	LAT:+00d00m00sLON:+000d00m00sALT:000mOBS:name	Setup → Observatory
IDCURSOR		
SAVEIMG	PNG/JPEG/BMP filename quality	File → Save Image...
PRINT	PRT/PS/BMP PORTRAIT/LANDSCAPE COLOR/BW filepath	Print or export to Postscript or Bitmap, File → Print and File→ Printer setup
SETNORTH		Chart → View Horizon → North
SETSOUTH		Chart → View Horizon → South
SETEAST		Chart → View Horizon → East
SETWEST		Chart → View Horizon → West
SETZENITH		z Icon (Right Bar)
ALLSKY		Icon (Right Bar)
REDRAW		Follows a modification command, mandatory to update chart
GETCURSOR		
GETEQGRID		
GETGRID		
GETSTARMODE		
GETNEBMODE		
GETAUTOSKY		
GETPROJ		

GETFOV	S/F	
GETRA	S/F	S → 17h07m12s F → 17.11991
GETDEC	S/F	
GETDATE		
GETOBS		
SETDATE	yyyy-mm-ddThh:mm:ss or "yyyy-mm-dd hh:mm:ss"	Setup → Date/Time
SETTZ	Etc/GMT	Setup → Observatory (Country Timezone)
GETTZ		
SETGRIDNUM	ON/OFF	Setup → Display → Lines - Show Grid Label
SETCONSTLINE	ON/OFF	Setup → Display → Lines - Show Constellation Figure
SETCONSTBOUNDARY	ON/OFF	Setup → Display → Lines - Show Constellation Boundary
RESIZE	width height	Resize the chart
GETRISESET		Get rise/transit/set time for the last selected object
MOVESCOPE	RA Dec [00.00]	Move the telescope cursor to coordinates. RA in decimal hours.
MOVESCOPEH	HourAngle Dec [00.00]	Same as MOVESCOPE but hourangle instead of RA. Hourangle in decimal hours.
IDCENTER		Identify object at chart center
IDSCOPE		Identify object at telescope cursor position
SHOWPICTURE	ON/OFF	Chart → Show objects → Show pictures
SHOWBGIMAGE	ON/OFF	Show background image
LOADBGIMAGE	fits_filename	Load new background image FITS file, also force reload of updated file if the name is the same.
LOADCIRCLE	file_name	Load a list of finder mark, same as right click - Finder circle - Load from file
SETCIRCLE	num diameter rotation offset	Define a finder circle as in Setup - Display - Finder circle
SETRECTANGLE	num width height rotation offset	Define a finder rectangle as in Setup - Display - Finder rectangle
SHOWCIRCLE	num_list	Set a comma separated list of circle to be activated 1,2,...,10
SHOWRECTANGLE	num_list	Set a comma separated list of rectangle to be activated 1,2, ...,10
MARKCENTER	ON/OFF	Show a mark at the chart center, same as Show mark
GETSCOPERADEC		Return current telescope coordinates.
TRACKTELESCOPE	ON/OFF	Same as menu Telescope / Track telescope.
CONNECTINDI		Connect to configured INDI telescope.
DISCONNECTINDI		Disconnect the INDI telescope.

SLEWINDI	RAhr Dec [in decimal]	Slew the INDI telescope to specified coordinates.
ABORTSLEWINDI		Abort the current slewing command.
SYNCINDI	RAhr Dec [in decimal]	Sync the INDI telescope at the specified coordinates.
CONNECTTELESCOPE		Connect to the default telescope.
DISCONNECTTELESCOPE		Disconnect the default telescope.
SLEW	RAhr Dec [in decimal]	Slew the default telescope to specified coordinates.
ABORTSLEW		Abort the current slewing command.
SYNC	RAhr Dec [in decimal]	Sync the default telescope at the specified coordinates.
OBSLISTLOAD	list_file_name	
OBSLISTFIRST		
OBSLISTLAST		
OBSLISTNEXT		
OBSLISTPREV		
OBSLISTLIMIT	ON/OFF	
OBSLISTAIRMASSLIMIT	[airmass]	
OBSLISTTRANSITLIMIT	[hours]	
OBSLISTTRANSITSIDE	EAST/WEST/BOTH	

V2.7 compatibility DDE command

Command	Parameters	Comment or GUI Equivalent
MOVE	obsolete RA: 00h00m00.00s DEC:+00d00m00.0s FOV:+00d00m00s	⚬ Icon
DATE	obsolete, same as SETDATE	
OBSL	obsolete, same as SETOBS	
RFSH	obsolete, same as REDRAW	
PDSS		Same as the menu Chart-Get DSS Image
SBMP	obsolete, use SAVEIMG	
SGIF	obsolete, use SAVEIMG	
SJPG	obsolete, use SAVEIMG	
IDXY	X:pixelx Y:pixely	
GOXY	X:pixelx Y:pixely	
ZOM+	obsolete, same as ZOOM+	
ZOM-	obsolete, same as ZOOM-	
STA+		⚬ icon (Main Bar)
STA-		⚬ icon (Main Bar)
NEB+		⚬ icon (Main Bar)
NEB-		⚬ icon (Main Bar)
GREQ	obsolete, use SETEQGRID	
GRAZ	obsolete, use SETGRID	
GRNM	obsolete, use SETGRIDNUM	
CONL	obsolete, use SETCONSTLINE	
CONB	obsolete, use SETCONSTBOUNDARY	
EQAZ	obsolete, use SETPROJ	

Directories and Files

Here is an overview of the directories and files on standard installations.

Windows

Purpose	Directory
executable	skychart.exe (the main program)
executable	cdcicon.exe (the clock with many times)
executable	varobs.exe (Variable Star Observer)
Installation	C:\Program Files\ciel
User data	C:\Users\[user]\AppData\Local\Skychart (Vista,Win7,Win8) C:\Documents and Settings\[user]\local Settings\Application Data\Skychart (Win XP)

Some directories may be hidden, change the Explorer setting to show them.

Linux

Purpose	Directory
executable	/usr/bin/skychart (the main program)
executable	/usr/bin/cdcicon (the clock with many times)
executable	/usr/bin/varobs (Variable Star Observer)
Installation	/usr/share/skychart
User data	~/.skychart (watch the preceding dot, this makes the directory 'hidden'.)

Mac OS X

Purpose	Directory
executable	skychart.app (the main program)
executable	varobs.app (Variable Star Observer)
Installation	/Applications/Cartes du Ciel
User data	~/Library/Application Support/skychart

The content of the user data directory

- **skychart.ini**. This file contains all configuration settings made by the user. Some examples: the observation position, the directories to the configured catalogs, and the definitions of your markers. In the releases before 2009-01-11, under Linux this file was known as .cartesduciel.ini in the user's home directory.
- **scope.ini** The settings for the internal LX200 or Encoder telescope driver.
- **cdc_trace.txt**. This file keeps track of the events with the last started instance of skychart.exe. It can be used for a simple form of debugging. Only found under Windows, under Linux this trace is to stdout.
- Archive A default location to save the FITS pictures.
- **database**. As the name suggests, this is the place where SkyCharts keeps a database with the name cdc.db that consists of indexed, searchable data. The database contains several tables. Herein the ephemeris of the asteroids and comets are kept after these were calculated from the orbital elements. Also there is a table that contains the ordered list of object-pictures that can be displayed. (For example, those in the data\pictures\sac subdirectory in the installation directory.) And the data for Countries and Observation locations are kept in this file.
- **MPC**. In this directory SkyChart saves the comet and asteroid data that was downloaded from the **Minor Planet Center [http://www.cfa.harvard.edu/iau/mpc.html]**.
- **pictures**. Directory where the FITS-pictures are saved, downloaded from the online DSS resources. Usually, you will find files with names like $temp.fit and $temp.fit.gz
- **satellites** The location to put your TLE files with the file extension .tle
- **script** Your personal tool box script go there.
- **tmp**. Directory where temporary pictures of planets are saved, retrieved from the **xplanet [http://xplanet.sourceforge.net/windows/]** program.
- **varobs** The Variable star observer catalog
- **vo** Your Virtual Observatory downloaded catalogs.

Script reference

This page contain reference material for scripting functions.

See the Tool Box description page for general information.

See the script example page for a quick start with the programming functions.

Script language

The language to use is Pascal Script [http://en.wikipedia.org/wiki/Pascal_Script].
For a complete reference of the Object Pascal language your can read the Free Pascal Reference guide [http://www.freepascal.org/docs-html/ref/ref.html]. But beware that some feature are not implemented by the script language, for example: no pointer, no assembler, no overloading.

In addition to the standard Pascal Script feature the following Skychart specific function are added.

Main menu function

Every items of the main menu can be used from the scripts.
The full list of menu items name is available from the menu list [http://sourceforge.net/p/skychart/code/HEAD/tree/trunk/skychart/menu.txt].

You can execute the menu action by using the Click method, or test if an option is activated by testing the Checked property.

Translation string

Every string translated for the main program can be used in a script.
Directly use the rsXXX constant to get the translated string.
The full list of available string is available from the source code [http://sourceforge.net/p/skychart/code/HEAD/tree/trunk/skychart/u_translation.pas].
If a string is not translated for your language don't hesitate to do it yourself.

Constants

name	value
deg2rad	degree to radian conversion constant
rad2deg	radian to degree conversion constant

Global variables access

function GetS(varname:string; var str: string):Boolean;	
Get the global string variable identified by varname	
varname	**value**
ChartName	The name of the last chart that send an event
RefreshText	The text of the ChartRefresh event
SelectionText	The short text of the selected object
DescriptionText	The object full description
DistanceText	The text of the last distance measurement event
Str1 .. Str10	Ten global variable for your use

function SetS(varname:string; str: string):Boolean;	
Set the global string variable identified by varname for later use	
varname	**value**
Str1 .. Str10	Ten global variable for your use

function GetSL(varname:string; var strl: Tstringlist):Boolean;	
Get the global stringlist variable identified by varname	
varname	**value**
Strl1 .. Strl10	Ten global variable for your use

function SetSL(varname:string; stlr: Tstringlist):Boolean;

Set the global stringlist variable identified by varname for later use	
varname	**value**
Strl1 .. Strl10	Ten global variable for your use

function GetI(varname:string; var i: Integer):Boolean;	
Get the global integer variable identified by varname	
varname	**value**
Int1 .. Int10	Ten global variable for your use

function SetI(varname:string; i: Integer):Boolean;	
Set the global integer variable identified by varname for later use	
varname	**value**
Int1 .. Int10	Ten global variable for your use

function GetD(varname:string; var x: double):boolean;	
Get the global double variable identified by varname	
varname	**value**
TelescopeRA	The telescope position right ascension
TelescopeDE	The telescope position declination
TimeNow	The current time in TDateTime format
Double1 .. Double10	Ten global variable for your use

function SetD(varname:string; x: Double):Boolean;	
Set the global double variable identified by varname for later use	
varname	**value**
Double1 .. Double10	Ten global variable for your use

function GetV(varname:string; var v: Variant):Boolean;	
Get the global variant variable identified by varname	
varname	**value**
Telescope1 , Telescope2	Two ASCOM Telescope objects
Dome1 , Dome2	Two ASCOM Dome objects
Camera1 , Camera2	Two ASCOM Camera objects
Focuser1 , Focuser2	Two ASCOM Focuser objects
Filter1 , Filter2	Two ASCOM Filter wheel objects
Rotator1 , Rotator2	Two ASCOM Rotator objects
Variant1 .. Variant10	Ten global variable for your use

function SetV(varname:string; v: Variant):Boolean;	
Set the global variant variable identified by varname for later use	
varname	**value**
Telescope1 , Telescope2	Two ASCOM Telescope objects
Dome1 , Dome2	Two ASCOM Dome objects
Camera1 , Camera2	Two ASCOM Camera objects
Focuser1 , Focuser2	Two ASCOM Focuser objects
Filter1 , Filter2	Two ASCOM Filter wheel objects
Rotator1 , Rotator2	Two ASCOM Rotator objects
Variant1 .. Variant10	Ten global variable for your use

Chart and Celestial objects

function Cmd(cname:string; arg:Tstringlist):string;
Execute one of the Skychart server command. Add the command name first to the string list, then each parameter.
procedure SendInfo(origin,str:string);
Send a message to the clients connected to the Skychart server.

function GetObservatoryList(list:TstringList):boolean;

Return the list of the Observatory favorite

function GetScopeRates(list:TstringList):boolean;

Return a list of speed rate supported by the telescope

function GetCometList(const filter: string; maxnum:integer; list:TstringList):boolean;

Return a list of comet according to the filter

function GetAsteroidList(const filter: string; maxnum:integer; list:TstringList):boolean;

Return a list of asteroid according to the filter

function CometMark(list:TstringList):boolean;

Mark the listed comet to the chart

function AsteroidMark(list:TstringList):boolean;

Mark the listed asteroid to the chart

Formating and conversion

Function ARtoStr(var ar: Double) : string;

Return a string formated Right Ascension of ar value

Function DEtoStr(var de: Double) : string;

Return a string formated Declination of de value

Function StrtoAR(str:string; var ar: Double) : boolean;

Convert a formated string to Right Ascension decimal value

Function StrtoDE(str:string; var de: Double) : boolean;

Convert a formated string to Declination decimal value

Function JDtoStr(var jd: Double) : string;

Format a julian date to YYYY-MM-DD string

Function StrtoJD(dt:string; var jdt: Double) : boolean;

Convert a formated string YYYY-MM-DD to julian date value

Function FormatFloat(Const Format : String; var Value : double) : String;

Format a decimal number according to the Format [http://www.freepascal.org/docs-html/rtl/sysutils/formatfloat.html] specification

Function Format(Const Fmt : String; const Args : Array of const) : String;

The Format [http://www.freepascal.org/docs-html/rtl/sysutils/format.html] Pascal function

function IsNumber(str: String): boolean;

Return True if the string represent a valid number

function StringReplace(str,s1,s2: String): string;

Replace all occurrence of s1 by s2 in str

procedure JsonToStringlist(jsontxt:string; var SK,SV: TStringList);

Parse a JSON formated string and return two stringlist. SK contain the names and SV the corresponding values

Dialog

function OpenDialog(var fn: string): boolean;

The standard Open File dialog. Return True if the OK button is pressed after the file selection.

function SaveDialog(var fn: string): boolean;

The standard Save File dialog. Return True if the OK button is pressed after the file selection.

function MsgBox(const aMsg: string):boolean;

A message confirmation dialog. Return True if YES is clicked.

Procedure ShowMessage(const aMsg: string);

Display a message.

function CalendarDialog(var dt: double): boolean;
The Skychart Calendar dialog. Return dt julian date

Run external program

function Run(cmdline:string):boolean;
Execute the specified command. Return immediately without waiting for the execution to end.

function RunOutput(cmdline:string; var output:TStringlist):boolean;
Execute the specified command, wait for termination and put the stdout to "output". **Beware** this function can completely lock the main program if it not finish in time.

function OpenFile(fn:string):boolean;
Open a document file using the default program

Also see the example about how to call a function in an external library.

TCP/IP client socket

For all this functions **socknum** identify the socket to use. This is a number between 1 and 10, thus allowing 10 simultaneous connection to different server.

function TcpConnect(socknum:integer; ipaddr,port,timeout:string):boolean;
Connect to the server at ipaddr:port and set the **timeout** for the subsequent operations

function TcpDisconnect(socknum:integer):boolean;
Disconnect from the server

Function TcpConnected(socknum:integer) : boolean;
Return **True** if the socket is connected

Function TcpRead(socknum:integer; var buf : string; termchar:string) : boolean;
Read data from the socket until the character **termchar** is encountered, typically termchar is CRLF

Function TcpReadCount(socknum:integer; var buf : string; var count : integer) : boolean;
Read data from the socket until **count** character are read or **timeout** is reached

Function TcpWrite(socknum:integer; var buf : string; var count : integer) : boolean;
Write data to the socket from **buf** for a length of **count**

Procedure TcpPurgeBuffer(socknum:integer);
Purge all the received data currently queued in the receive buffer

Script example

This page give tips and example of scripting functions.
You can also look at the three standard tool box code from within the program.
For more details about a specific function see the script reference page.

Generality

We first look in detail at the code of the Goto button of the Observer tools standard tool box.
This cover many programming basis.

The full script code look as following:

```
// Slew telescope
  var ra,de: double;
      a,b,r: string;
      c: Tstringlist;
  begin
  memo_1.clear;
  if MenuTelescopeConnect.checked then begin
     if not StrToAR(Edit_1.text,ra) then begin memo_1.lines.add(rserror+' RA!');exit;end;
     if not StrToDE(Edit_2.text,de) then begin memo_1.lines.add(rserror+' DE!');exit;end;
     a:=floattostr(ra);
     b:=floattostr(de);
     GetSL('STRL1',c);
     c.clear;
     c.add('SLEW');
     c.add(a);
     c.add(b);
     r:=cmd('',c);
     c.clear;
     memo_1.lines.add(r);
  end
  else memo_1.lines.add(rsTelescopeNot);
  end.
```

Take a look at each part in detail:

```
// Slew telescope
```

Is a comment, you can use // {..} (*..*) to enclose your comments.

```
  var ra,de: double;
      a,b,r: string;
      c: Tstringlist;
```

Define the variable we use later in the script.
Important variable type are: integer, double, string.
The Tstringlist type is use here to send a command to Skychart.

```
  begin
```

The start of our program.

```
memo_1.clear;
```

Clear the text box we use to show the messages. This ensure the text box is not filled by previous messages.

```
  if MenuTelescopeConnect.checked then begin
```

We test the Checked property of the menu item MenuTelescopeConnect. This indicate we have connected the telescope to the program.
If the result is true we execute the code block starting at "begin" up to the corresponding "end"

```
     if not StrToAR(Edit_1.text,ra) then begin memo_1.lines.add(rserror+' RA!');exit;end;
```

We try to convert the RA in HMS format from the text in Edit_1 text box to a numeric value. If the conversion fail (because we type some junk in the text box) we show an error message and exit.

```
     a:=floattostr(ra);
```

Convert the RA back to string representation but with the decimal format required by the command.

```
GetSL('STRL1',c);
c.clear;
```

Request a TStringList object identified by STRL1. We clear any data that may stay in the object.

```
c.add('SLEW');
c.add(a);
c.add(b);
```

Add the command and the required parameters (in this case RA and DEC) to the stringlist.

```
r:=cmd('',c);
```

Execute the command and store the result in the variable r.

```
memo_1.lines.add(r);
```

Show the result of the command to the text box.

```
else memo_1.lines.add(rsTelescopeNot);
```

The case the test MenuTelescopeConnect.checked is false we execute this line.
It show in the text box a translation in the local language of 'Telescope not connected'.

```
end.
```

The end of the program.

Call an external library

You can define a function in an external library for use within your script as another local function.

This example implement a simple chronometer by using the GetTickCount function of the Windows API.
There is two button Start and Stop and two text box. A global integer variable is used to store the start time.

Script for the Start button:

```
function GetTickCount: Longint; external 'GetTickCount@kernel32.dll stdcall';
var
  tick: Longint;
begin
  tick:=GetTickCount;
  setI('Int1',tick);
  edit_1.text:='Started';
end.
```

Script for the Stop button:

```
function GetTickCount: Longint; external 'GetTickCount@kernel32.dll stdcall';
var
  t: double;
  t1,tick: Longint;
begin
  tick:=GetTickCount;
  getI('Int1',t1);
  t:=double(tick-t1)/1000;
  edit_1.text:='';
  edit_2.text:=formatdatetime('HH:MM:SS.ZZZ',t/24/3600);
end.
```

You can call any library function this way but beware this is system dependent, the kernel32.dll library is not available on Mac or Linux.

Another limitation is that many library function expect a pointer to a parameter structure. As the script language use byte code internally (as Java) it cannot use a pointer to give the parameters. A solution is to write a C library wrapper that export the function with a flat parameter list.

Using ASCOM directly

This describe how to use an ASCOM device directly from your script without any use of the Skychart internal ASCOM telescope.

This can be use to access another class of device, the example here connect to a dome, or to access additional properties for your telescope.
In the later case you must be careful that your script work as a concurrent to Skychart main program for the device access.

Use the ASCOM chooser

The following code assigned to a button allow to select the ASCOM Dome driver we want to use. The driver name is saved in the text field Edit_1.

```
var
```

```
  V: variant;
  w,s: widestring;
begin
  V := CreateOleObject('ASCOM.Utilities.Chooser');
  w:='Dome';
  V.DeviceType:=w;
  s:=edit_1.text;
  s:=V.Choose(s);
  edit_1.text:=s;
  V:=Unassigned;
end.
```

Replace w:='Dome'; by Telescope, Focuser, Rotator, Camera, Filter to select another driver class.

Connect to the ASCOM driver

The following code is for the "Connect" button. It connect to the ASCOM Dome driver we select previously. We use the global variable Dome1 to store the ASCOM object.

```
var
  D: variant;
  s: widestring;
begin
  s:=edit_1.text;
  getV('Dome1',D);
  if VarIsEmpty(D) then
    D := CreateOleObject(s);
  D.connected:=true;
  setV('Dome1',D);
end.
```

Use the ASCOM driver

Now we want to add a button to open the dome shutter. This is just an example, at this point any ASCOM property can be use. The first test protect again a program crash if we try to use an initialized variant. The second test protect again an ASCOM error if the dome is not connected.

```
var
  D: variant;
begin
  getV('Dome1',D);
  if (not VarIsEmpty(D)) then
    if D.Connected then
      D.OpenShutter;
end.
```

Disconnect the ASCOM driver

Add a button to disconnect the driver and release the resources.

```
var
  D: variant;
begin
  getV('Dome1',D);
  if (not VarIsEmpty(D)) then
    if D.Connected then
      D.connected:=false;
  D:=Unassigned;
  setV('Dome1',D);
end.
```

Open a document

The following code open the Skychart documentation page in the default web browser.
You can use any document type with this function, the document open with the default application the same way as if you double click the document in the file explorer.

```
begin
  OpenFile('doc\wiki_doc\en\documentation\start.html');
end.
```

Run a command

There is two different way to run an external command or program, depending if you want to wait for a result or not.

Wait for a result

The following command run the DIR command in the current directory. The result is stored in a stringlist and later show in a text memo. It contain the list of files in the directory.

```
var r:TstringList;
begin
  GetSL('STRL1',r);
  r.clear;
```

```
  RunOutput('dir',r);
  Memo_1.lines.assign(r);
end.
```

No wait

If the command can run for an undetermined time or do not produce an output you need to use the following form. This example run the Variable star observer program and exit immediately.

```
begin
  Run('varobs');
end.
```

Computational method and precision

This page give some information about the computation method used by *Cartes du Ciel - Skychart* and the precision you can expect for the displayed values.

You must be careful this description is valid with the standard configuration setting of the program, using the default catalog data. You have many option available to alter the results, principally in the <u>Chart,Coordinate</u> setting page. Use this settings only if you really know what you do!

Stellar equatorial positions

The basic precision depend on the star catalog used, for the precision of the position but also for the proper motion. The default catalog is the Extended Hipparcos Compilation (<u>XHIP, V/137 [http://cdsarc.u-strasbg.fr/viz-bin/Cat?V/137D]</u>). The advantage of this catalog is the availability of the full space proper motion parameters for almost every stars.

After it get the catalog data the program compute the position corrected for the star proper motion at the current chart date using the pmRA and pmDEC values, and full space motion if the parallax and radial velocity are available (u_projection.pas, ProperMotion). This give the equatorial astrometric J2000 position.

Then the precession is computed for the chart date using the method given by J. Vondrak, N. Capitaine, P. Wallace in "<u>New precession expressions, valid for long time intervals A&A 2011 [http://adsabs.harvard.edu/abs/2011A%26A...534A..22V]</u>" (u_projection.pas, PrecessionV). This give the equatorial mean of date position.

To find the apparent position we compute the nutation using the value given by the JPL ephemeris, then the annual aberration and the light deflection by the Sun (u_projection.pas, apparent_equatorialV). This give equatorial apparent position.

For the current epoch the precision is expected to be better than 0.1 arcsec.
The precision of the proper motion calculation over a long time period depend of the availability of the parallax and the radial velocity, but also of the standard error on the values. An error of about 1 arcsecond by millennium is to be expect.
The precession computation is valid for a +/- 200'000 years period. The precision is better than one milli arcsec at the current epoch, it reach a few arcseconds throughout the historical period, and a few tenths of a degree at the end of the period.

DSO equatorial positions

The main problem about the position of deep sky objects is the difficulty to precisely define the center of the object. Because of this difficulty the position differ when using different source catalog. Also many historical catalog still in use give the position with one arcminute precision only.

After it get the catalog data the program compute the precession and the apparent position as describe above for the stars.

Planets equatorial positions

The position of the planets are computed using the JPL ephemeris or if no file are found for the current date, the library <u>plan404 by Steve Moshier [http://www.moshier.net/]</u> that allows for computation from -3000 to +3000 with a precision better than one arc second.
By default an extract of DE405 valid between 2000 and 2050 is supplied with the program. So the first thing to do if you want long term high precision planet position is to install a <u>full DExxx [ftp://ssd.jpl.nasa.gov/pub/eph/planets/Linux/]</u> file.
DE431 is recommended if you can afford the 2.5GB download. With this file you can compute precise planet position and nutation between -13000 and +17000.

The computation function return the J2000 planet position corrected for light time, so the program use the same function as for the stars to compute the precession for the current date. This is the geocentric mean of date position.

Then we correct the position for the parallax for the observer location on Earth (u_projection.pas, Paralaxe). This give the topocentric mean of date position.

Then the apparent position is computed by applying the nutation and annual aberration (not for the Moon). This is the topocentric apparent position.

For the current epoch the precision is expected to be better than 0.1 arcsec.
For a date far in the past or the future the major source of error is the uncertainty in the difference between the universal time and the terrestrial time <u>deltaT [http://en.wikipedia.org/wiki/%CE%94T]</u>. You can see and change the value of deltaT in the <u>Time setting</u> window.
The precision of the computation itself depend on the individual ephemeris, but it is always far better than every expectation for a terrestrial observer. Refer to the JPL documentation.
The error on precession is the same as discussed for the stars.

Comets and asteroids equatorial positions

Comets and asteroids computation are based on elements in MPCORB format. You need to <u>download</u> the required elements first.

The elements are then loaded in a database that allow for many set valid at different epoch. The program always use the element set the nearest to the current date.

For the asteroids it also compute a monthly value of the magnitude that is used to exclude the objects that are currently way to faint to be visible. This help to speed up the other computation.

When the current day change, the program compute a position for each object. This position is then used to know if a precise position need to be computed for the current chart FOV. The NEO are exclude from this process because the position change too rapidly. All of this processing is require to avoid to compute too much position every time the chart is refreshed.

After the elements for an object are selected, the program compute the heliocentric rectangular coordinates and then the J2000 geocentric position corrected for light time.

Then precession, parallax and apparent position is computed the same as for the planets.

When using current element data the precision is expected to be about 0.1 arcsec.

You can reliably compute the asteroids and comets position only for a few month around the date of the elements. So it make no sens to compute this position for a date far in the past or future.

Alt/Az positions

This is how the program convert the apparent equatorial position of any object to the azimuth/altitude at a given location.

We first get the geometric azimuth/altitude by a rotation of the coordinate system using the equatorial coordinates, the sidereal time and the observer latitude.

If you give the current Earth pole coordinates in the Observatory settings the position is corrected for this small offset.

Then the position is corrected for the diurnal aberration and refraction.

The refraction is computed using two different method, one for the display on the map, the other to display a more precise value in the detailed information window.

The first method need to be fully reversible without too much computation. It is currently based on Bennett formula.

The second is based on the method in SLALIB [http://star-www.rl.ac.uk/docs/sun67.htx/sun67.html] (REFCO,REFZ,REFRO) and take account for more atmospheric parameters. To fully benefit of this increased precision you need to carefully indicate the atmospheric pressure, the temperature, the relative humidity and if possible the tropospheric rate (from a nearby sounding or a meteorological model). The wavelength used for the computation is 550nm.

If all the observatory parameters are given with the maximum precision, the precision of the azimuth and the geometric altitude must be better than 0.5 arcsec. The precision on the refracted altitude depend on the difference between the model and the real atmosphere.

But remember that 0.1 arcsec represent 3 meters on the soil and a star on the celestial equator move by this distance in 0.007 second. You need to set your observatory location and measure the time with this precision if you want it make some sens.

How to install the software from the source code

Get the source code

Create a new directory for the source code. You have two ways to get the code:

- Download the source file skychart_v3_xxx_source.tar.gz
  to this directory and extract the files.
  On Linux use : tar xzf skychart_v3_xxx_source.tar.gz
  On Windows use 7-zip available from: http://sourceforge.net/projects/sevenzip/ [http://sourceforge.net/projects/sevenzip/]

  - Or better use a Subversion client [http://subversion.tigris.org] to ease the update of daily changes.
    The command is:

```
svn checkout svn://svn.code.sf.net/p/skychart/code/trunk .
```

Automated Compilation and installation

If you just want to compile the software without using the Lazarus interactive environment you can use the scripts available in the base directory.

Before running these scripts be sure to have the Free Pascal binaries in your PATH environment, this is where you find the fpcmake command.

In the base directory you also found the daily_build.sh script I use to automatically build the packages for Linux and Windows.

For Linux and Mac

```
./configure [fpc=free_pascal_path] [lazarus=lazarus_path] [prefix=installation_path]
make
make install
make install_data
```

For example to install the complete software from source on Ubuntu 11.10:

```
sudo apt-get install build-essential lazarus subversion
svn co https://skychart.svn.sourceforge.net/svnroot/skychart/trunk skychart
cd skychart
./configure fpc=/usr/lib/fpc/2.4.4 lazarus=/usr/lib/lazarus/0.9.30 prefix=/usr/local
make
sudo make install
sudo make install_data
sudo make install_cat1
sudo make install_cat2
sudo make install_pict
```

For Windows

- Beware not to have another make command than the Free Pascal one in your path.
- Manually compile the library getdss and plan404 with Mingw [http://www.mingw.org/].
- Install the command sed for Windows [http://gnuwin32.sourceforge.net/packages/sed.htm].
- Edit the file configure.cmd to adjust the values for sed=, fpc=, lazarus=, prefix=
- You may have to adjust the scripts according to the Linux version because I not use them and they are probably out of date.

```
configure.cmd
make
make install
make install_data
```

Interactive compilation

First, install the required components from src/skychart/component directory.

Click "Open Package", select "component/cdccomponents.lpk", click "Compile", "Install".
When the install tells you to rebuild Lazarus say Yes.

You can now open the main project files skychart/cdc.lpi and compile.

Lazarus show by default the last form added to the project. To view the main form use the menu Project / Forms, and select f_main.

To run in debug adjust Run-Run Parameters-Working Directory to your CDC directory.

To reduce the executable size for production, use strip and upx.

The library getdss and plan404 are written in C language. To compile them, install the gcc compiler (Mingw [http://www.mingw.org/] on Windows) and run make from each library folder.

A few Windows specific libraries and plugins are not yet ported to Lazarus, please use Delphi if you want to compile them.

Install Lazarus

To know which version of Lazarus is require for a specific version of Skychart, install the binary version and look at the menu Help / About. There is a line that show the FPC and Lazarus version used.

Install Lazarus from http://lazarus.freepascal.org [http://lazarus.freepascal.org]. See http://wiki.lazarus.freepascal.org/Installing_Lazarus [http://wiki.lazarus.freepascal.org/Installing_Lazarus] for more information.

Launch Lazarus and open the menu Package - Install/Uninstall Packages.

Check that Printer4Lazarus and TurboPowerIPro are installed, this is normally the case.
If not, install them from lazarus/component :

- printers/printer4lazarus.lpk
- turbopower_ipro/turbopoweripro.lpk

I use the following procedure to install or update Free Pascal and Lazarus on Linux with the cross compiler for Windows:

```
cd ~/fpc
# svn co http://svn.freepascal.org/svn/fpc/branches/fixes_2_2 .
svn up
make clean
make build
sudo make install
make clean OS_TARGET=win32 CPU_TARGET=i386
make build OS_TARGET=win32 CPU_TARGET=i386
sudo make crossinstall OS_TARGET=win32 CPU_TARGET=i386
ver=`fpc -iV`
sudo ln -f -s /usr/local/lib/fpc/$ver/ppc386 /usr/local/bin
sudo ln -f -s /usr/local/lib/fpc/$ver/ppcross386 /usr/local/bin
cd ~/lazarus
# svn co http://svn.freepascal.org/svn/lazarus/trunk .
svn up
make clean
make OS_TARGET=win32 CPU_TARGET=i386 clean
make bigide
make OS_TARGET=win32 CPU_TARGET=i386 bigide
```

Naming convention for the program source

The following naming convention is used for the main project source code to enable quick recognition of the destination of the units.

```
cdc.lpi        : The main project
pu_*.pas       : Form units with specific code only
pu_*.lfm       : Form definition
cu_*.pas       : Unit containing non-visual object.
u_*.pas        : Unit with generic code.
```

Directory structure

```
|- src -|                     < base directory, compilation scripts
        |- skychart |         < skychart module
                    |- component   < project component
                    |- library     < project library
                    |- ...
                    |- units       < all compilation object go here

        |- varobs   |         < varobs module

        |- tools | - data     < the "data" directory structure require to run the progran
                 | - cat      < the basic catalogs, the program used to build them
                 | - ...      < other data files
```

FAQ

General

⑦ **How can I contribute to *Cartes du Ciel-SkyChart*?**

You can contribute in several ways:

- Try the last development version and report the bugs you find to the bug tracker.
- Volunteer to write copy, or edit this text to solve spelling and grammatical errors.
- Translate the program, or this web text into your native language.
- Review the program code and propose your own improvements.
- Please send me your ideas [http://www.ap-i.net/mantis/], I am interested in your suggestions.

⑦ **So version 3.10 is more recent that 3.8 ?**

Yes, the version number is not a decimal number but a combination of two number separated by a dot.

The first number "3" is the major version, it normally change only if the new version is incompatible with the previous one. The last time it occur was from 2 to 3 in 2010. A new change is not planned for now.

The second number "8" or "10" is the minor version, it change at each release of a stable version. An even number indicate a stable version. The odd minor numbers are used for the development and the intermediate beta version.

Installation

⑦ **I have *Cartes du Ciel-SkyChart* version 2.76. What do I need to do to run the new version 3? How many files do I download and do I delete 2.76 first?**

You just need to install the version 3.0 for Windows in the same directory as the **version 2.76**. Do not try to install in another directory, and do not delete anything because you can use the two versions at the same time. This is useful to validate that the V3 work fine for you. Run cdc.exe for the V3.0 or ciel.exe for the V2.76

⑦ **Are the earlier catalogs associated with CdC V2.76 compatible with *Cartes du Ciel-SkyChart* V3 ?**

Yes, all these catalogs can be used with the version 3.0. The only exception is the external catalog, they are replaced by the more convenient Catgen text catalog or the Virtual Observatory catalog.

⑦ **Where do I install the extra catalogues under Linux, in which directory do they go into?**

If you have root permissions, it is best to put them in /usr/share/skychart/cat to have them with the basic catalogs. But you can also install them anywhere as long as you specify the full path in the catalog configuration menu.

⑦ **Can I get *Cartes du Ciel-SkyChart* software for a Macintosh?**

Yes, the version 3 now work the same on Mac OS X than on Windows or Linux. Look at the specific installation instruction for more information about the additional software you need to pilot a telescope or compute the artificial satellites.

⑦ **What can I do with MySQL?**

MySQL was used as the default database up to the version alpha 7. Since June 2005 it is replaced by the more simple SQLite. You can still use MySQL if required, for example to share the database across a local network. The database is used to store the following data: asteroids and comets elements, object pictures and world locations.

Usage

⑦ **Does *Cartes du Ciel-SkyChart* V3.0 for Mac/Linux support telescope tracking?**

Yes, *Cartes du Ciel-SkyChart* V3.0 can use **INDI [http://indi.sourceforge.net/]** to interface the telescope. Any telescope model supported by INDI must work with *Cartes du Ciel-SkyChart*. Also you can use the include interface for Meade/LX200 protocol or a simple encoder interface.
The **ASCOM [http://ascom-standards.org/]** standard is specific for Windows and can be run only with this platform.

⑦ **There is no "follow" and "see" telescope in the telescope menu of the V3.0?**

There is now a **Track telescope** entry in the Telescope menu
You can also use the "lock on" feature (the anchor button). When the telescope is connected and no other object is selected, this button locks on the telescope.

⑦ **Why is the sky blue?**

Because it is daytime ☺ or the Moon is up. You can change that from the configuration menu, **Display → Sky colour**. Or change the date / time in the program by **Setup → Time**. Or change the coordinates system from Alt/Az to one of the other systems by a click on one of the icons on the left bar in the **Coordinates system group**

How can I see a sky chart with simulation of the planets before 3000 B.C?

Since the version 3.9 you can use the JPL ephemeris DE431 that extend the computation range for the planet to -13000 → +17000.

How can I switch the language from English to my language?

The language is normally detected automatically when you start the program. The menu **Setup → General → Language** gives you the opportunity to select another language.

Does *Cartes du Ciel-SkyChart* V3 have DDE support?

No, the standard communication protocol for the V3 is TCP/IP for portability reasons.

Why are the the fonts with my Linux version too small / too big?

There was a problem about Gtk1 font settings on Linux in the past. In the latest versions of Skychart, Gtk2 is used by default. There should not be such a problem now. If so, please report the bug to us.

Does *Cartes du Ciel-SkyChart* V3.10 for Windows support Windows 10?

Yes - unofficially this has been tested against OS Version: 10.0.10240 N/A Build 10240 on 27-10-2015

Common Problems

The 64bit version of the program crash on startup

You need an AMD or Intel 64bit processor more recent than 2005 or 2006 that support the LAHF and SAHF instructions. Otherwise you must use the 32bit version.
See: https://en.wikipedia.org/wiki/X86-64#Older_implementations [https://en.wikipedia.org/wiki/X86-64#Older_implementations]

On Linux, every character I type in the search box are doubled

This is a bug with some input method. The solution is to install the ibus-gtk package for your system and set ibus as your default input method.
Or if it is not possible to change the default, run skychart with the following command:

```
GTK_IM_MODULE=ibus skychart
```

I've installed CdC V3.0 but program doesn't display stars or objects.

Maybe you have only installed the program. You also have to install the base catalog package, it is normally include with the standard download. Look at **Download** page and follow instructions in the **Installation of extra catalogs**.
Are your star catalogs activated? Make sure they are, see the **Stars** tab from the **Setup → Catalog** dialog. Another possibility is that you accidentally switched the display of stars **off**. Check the possibility of the ⁘ icon in the **object group A** of the object bar.
Or maybe the cause is hidden in a troubled display filter. Check your settings in the tab **Object Filter** from the **Setup → Chart, Coordinates** dialog.

The help do not open, instead there is a message 'Unable to locate HTML Browser'.

This is because of a misconfiguration of your Windows system regarding the default browser.

Take the following action in sequence:

- Be sure you have a web browser installed and fully functional on your computer.

- From the Windows Start menu select "Default Programs", in the new window select "Set program access and computer defaults", select your preferred program for "default web browser".

- From the Windows Start menu select "All programs", Accessories", "Command prompt". In the command line window type

  ```
  path
  ```

  and press enter key. Check that C:\Windows\system32 is include in the result line.
  If not do the following:

  - Open the Windows Control Panel, select "System and security", "System". On the left pane select "Advanced system settings". Click the button "Environment Variables…".
    Look for a PATH variable in both the user and system variables. Select the row, click the Edit button and add "C:\Windows\system32;" at the beginning of the row "Variable value".
    You may need to restart your computer to be sure the change take effect.

- Return to the command prompt window.
  Type (copy/paste) this command:

  ```
  rundll32.exe url.dll,FileProtocolHandler "C:\Program Files\Ciel\doc\wiki_doc\en\documentation\start.html"
  ```

Your web browser must open the help start page. If not, read any error message that can help to investigate further.

Documentation License

Software License

Cartes du Ciel software, Copyright © 2006 Patrick Chevalley

Table of Content